Digital Image Processing using SCILAB

Rohit M. Thanki • Ashish M. Kothari

Digital Image Processing using SCILAB

Springer

Rohit M. Thanki
C. U. Shah University
Wadhwan City, Gujarat, India

Ashish M. Kothari
E.C. Department
Atmiya Institute of Technology and Science
Rajkot, Gujarat, India

ISBN 978-3-030-07795-2 ISBN 978-3-319-89533-8 (eBook)
https://doi.org/10.1007/978-3-319-89533-8

Printed on acid-free paper

This Springer imprint is published by the registered company Springer International Publishing AG part of Springer Nature.
The registered company address is: Gewerbestrasse 11, 6330 Cham, Switzerland

Preface

The ultimate question that arises for us is, what happens when we look at an object? The process starts with the eye capturing the object and sending signals to the brain. These signals are decoded by the brain and important information is obtained—this is called *image processing*. This combination of the eye and brain develops the ultimate imaging system. Image processing as a subject involves processing images (captured by camera) by computers using algorithms written for this purpose. Here, the camera replaces the eye and the computer does the work of the brain.

Image processing involves changing the nature of an image in order to either improve its pictorial information for human interpretation or render it more suitable for autonomous machine perception.

This book contains the fundamentals of image processing techniques and various ways to modify image data to achieve the desired image data. This book also introduces the concepts and fundamentals of SCILAB—an open-source tool for processing discrete time signals. With the help of SVIP Toolbox, image data can also be processed and modified using SCILAB. This book begins with the fundamentals of SCILAB and a basic introduction to it. It presents various kinds of signals, from one-dimensional to multidimensional, with a specific focus on two-dimensional image signals. The book delves into the importance of image enhancement techniques and explains the fundamentals of the histogram and its processing. In addition, filtering an image's data signal is considered specifically important. Since noise plays a very important role in image processing, this book focuses on restoration of the image from a degraded version that has occurred because of the introduction of noise. An explanation of various image segmentation techniques is also provided, followed by a detailed discussion of one very important aspect of image processing: morphological image processing. This is followed by discussion of some aspects of color image processing. Finally, some real-world applications of image processing are discussed.

Wadhwan City, Gujarat, India Rohit M. Thanki
Rajkot, Gujarat, India Ashish M. Kothari

Acknowledgments

We express our sincere gratitude to Rafael C. Gonzalez and Richard E. Woods for their deep understanding of the basic theories of image processing. We also express deep appreciation and indebtedness to the University of South Carolina for providing access to their public domain image database for implementation of SCILAB codes for image processing theories.

We also extend many thanks to the staff of the Atmiya Institute of Technology and Science and C. U. Shah University for their cooperation and friendship. We are also grateful to both our family members and friends for their moral support.

Contents

List of Figures

Chapter 1
Introduction

1.1 Introduction

An image is a type of signal. A signal is a function that carries information regarding physical system behavior or some of the characteristics of a system. A simple example is the color indicators at a traffic signal about the status of traffic. These indicators indicate different characteristics such as green for walk, yellow for get ready to drive, and red for stop. The signal can be represented in many ways that contain information in various patterns. In an electrical system, current and voltage are presented as signals and called electrical signals. These signals are widely used in radio transmission, mobile communications, medical applications, and military applications such as radar, sonar, and in many other areas where electricity is required. Human speech or music is known as acoustic wave signals. The human speech signal depends on the speaking energy of humans whereas the music signal depends on energy generated by various musical instruments.

A signal has a duration that varies and is represented by a pattern in the spatial domain. Mathematically, signals are represented as functions of one or more independent variables that can be either continuous or discrete. The continuous time signal is represented by a function that continues the time interval. A discrete time signal is represented by a function some instants of time interval. Signals with continuous time and continuous amplitude are called analog signals. These signals can be converted into discrete or digital forms using various digital devices.

Signal processing is the process of extraction of important information from the signal. Digital signal processing (DSP) is related to the representation of signals discrete in nature. It was invented in the seventeenth century and has become an important tool for modern science and technology. DSP can be divided into various sub-areas such as digital image processing, statistical signal processing, communications, biometrical signal processing, speech signal processing, spectral estimation and military signal processing, etc.

R. M. Thanki, A. M. Kothari, *Digital Image Processing using SCILAB*,
https://doi.org/10.1007/978-3-319-89533-8_1

A human is a natural signal processing system. Humans see objects using a visual system and hear signals using an auditory system. The human visual system (HVS) plays a key role in navigation, identification, verification, gait, gesture, posture, communication, psychological interpretation, etc. The human auditory system (HAS) converts sound waves into nerve impulses, to analyze auditory events, remember and recognize sound sources, and perceive acoustic sequences.

1.2 Types of Signals

In the world, many kinds of signals are presented with their own physical significance. Many types of signals are sensed by humans who interpret or decode a signal for their own purpose. The signal can be characterized based on its shapes, energy, time intervals, and other physical properties. Based on a time interval, a signal can be classified as a continuous time signal or a discrete time signal. Based on the amplitude and the time axis, the signal can be classified as an analog signal or a digital signal. If the signal repeats in some time interval, they are called periodic signals; otherwise it is called as an aperiodic or nonperiodic signal. If each value of a signal is defined, then it is called a deterministic signal; otherwise, it is called as a random signal. In the classification based on dimensionality, signals are classified as one-dimensional (1D), two-dimensional (2D), three dimensional (3D), and multi-dimensional signals. These signals are explained below.

1.2.1 One-Dimensional Signals

A 1D signal is usually treated as a time waveform, i.e., $x(t)$. The processing of this signal has a sound history and is of great importance in various fields as speech communication, biomedical engineering, military communication. An example of a 1D signal is shown in Fig. 1.1. For example, when a user is employs a mobile telephone for communication, then his/her voice is converted in to an electrical signal and transmitted to another place using a communication system.

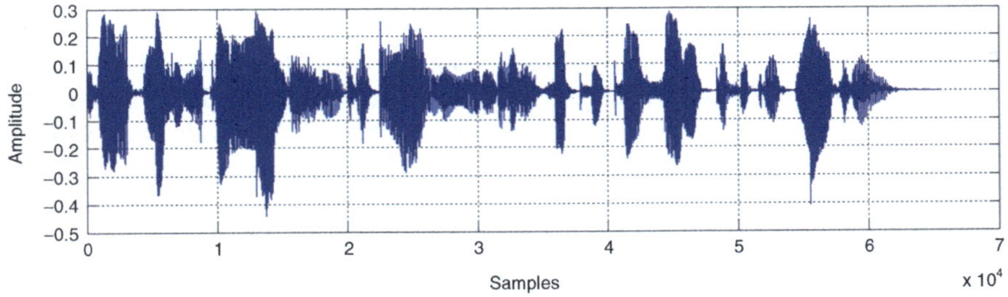

Fig. 1.1 Example of a 1D signal

1.2.2 Two-Dimensional Signals

A 2D signal is a function of two independent variables, i.e., $f(x, y)$. This signal gives functional behavior in terms of energy distributed over the (x, y) plane. An example of a 2D signal is a digital image. Everyday scenes observed by a human are a combination of various illuminated objects. The light energy reflected from these objects can be considered a 2D intensity function, which is commonly known as an image.

1.2.3 Three-Dimensional Signals

Image of still objects is a function of the (x, y) plane, By adding a third element as a time interval, the 3D signals can be obtained which represents image sequences of a dynamic scene and are called video signals. An example of a 3D signal is an image sequence obtained from satellite sensors (Fig. 1.3).

1.2.4 Multi-Dimensional Signals

When a signal is described by more than one variable or function, then it is called a multi-dimensional signal. As discussed in previous subsections, an image as a 2D signal and a video as a 3D signal are examples of multi-dimensional signals. This signal processing is an innovative field of interest in an industry where information is acquired from various dimensions. Some of the applications of this signal processing are 3D face modeling, 3D project development, and 3D object tracking.

Fig. 1.2 Example of a 2D signal

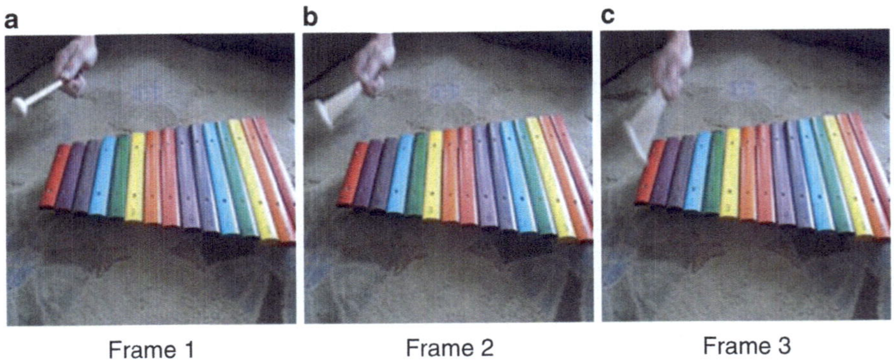

Fig. 1.3 Example of a 3D signal. (**a**) Frame 1, (**b**) frame 2, (**c**) frame 3

1.3 A Digital Image and Its Processing

A digital image is a function of a finite number of elements, each of which has a particular location and value [1]. These elements are known as picture elements, image elements, and pixels. The term pixel is widely used to represent a digital image. The first question that arises in our minds is "What happens when we look at an object?" The process starts with the eye capturing the object and sending signals to the brain. These signals are decoded by the brain and valuable information is obtained. This is called image processing.

We start processing images from the first day of life and we continue doing so until the last moment; thus, we can say that it is an integral part of our day to day lives. Therefore, the combination of eye and brain develops an ultimate imaging system. Image processing as a subject involves processing images (captured by a camera) by computers where algorithms are written to process images. Here, the camera replaces the eye and the computer does the work of the brain. Image processing involves changing the nature of an image to either:

1. Improve its pictorial information for human interpretation, or
2. Render it more suitable for autonomous machine perception.

Today, image processing is used everywhere in the world. Some of the applications are listed in this section. The areas of applications of image processing are classified according to the images generated from their energy source. The principal energy source for images today is the electromagnetic energy spectrum, and other energy sources may be acoustic, ultrasonic, and electronic [1]. Figure 1.4 shows the electromagnetic energy spectrum.

Images generated using a gamma-ray are called gamma-ray imaging. Those generated using X-rays are X-ray imaging. These types of images are widely used in the inspection of the human body in medical science. The major field of application for gamma ray imaging is nuclear medicine and astronomical observation. Figure 1.5 shows imaging in gamma ray bands and X-ray bands [2].

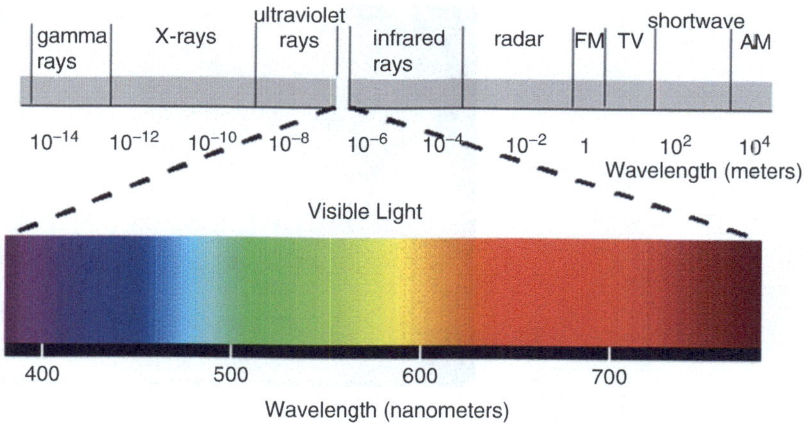

Fig. 1.4 Electromagnetic energy spectrum

Fig. 1.5 Example of (**a**) gamma-ray imaging, (**b**) X-ray imaging

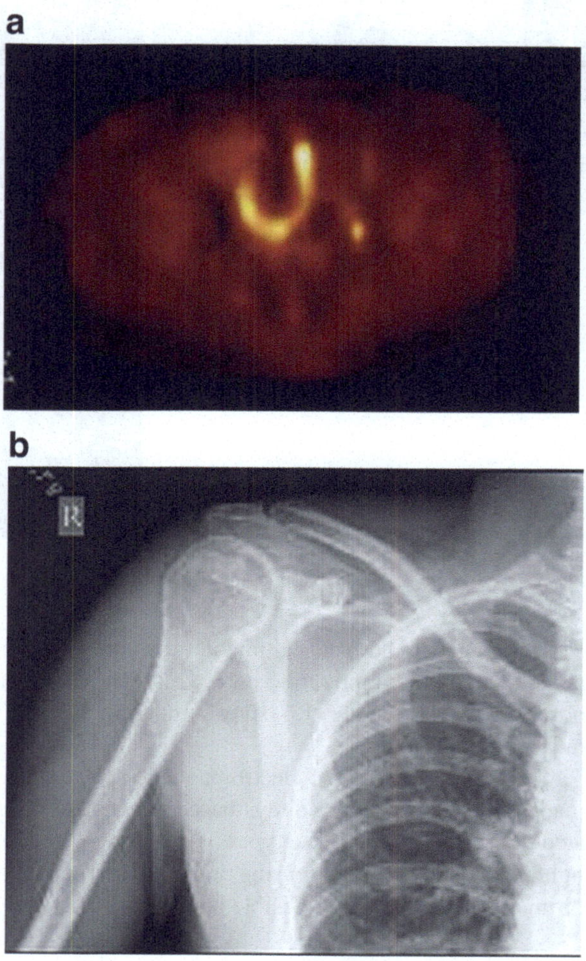

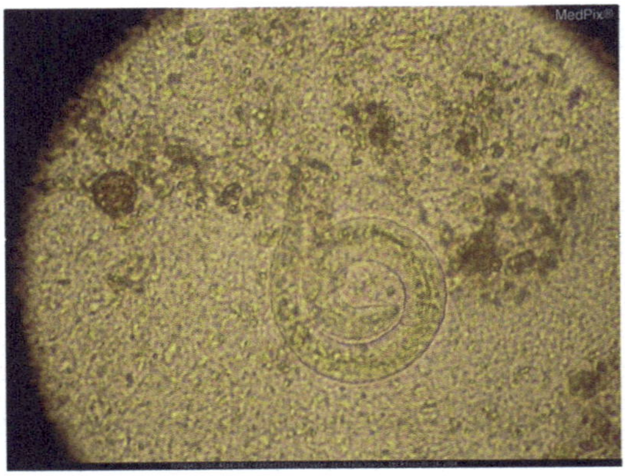

Fig. 1.6 Example of ultraviolet imaging

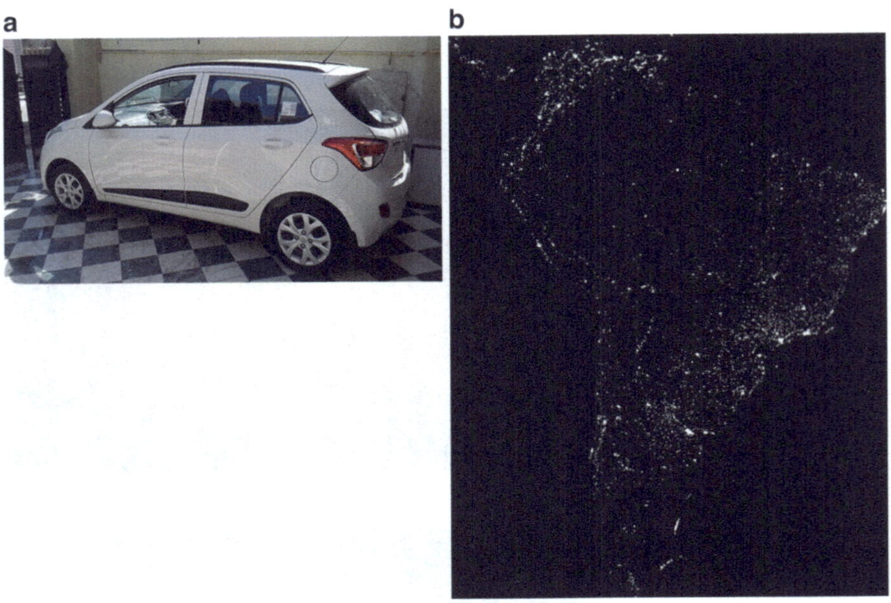

Fig. 1.7 Example of (**a**) visible imaging, (**b**) infrared imaging

Images generated using ultraviolet rays are called ultraviolet imaging. Applications include lithography, microscopy, lasers, biological imaging, and astronomical observation [1]. Figure 1.6 shows imaging in this band, with an example of microscopy [2].

The visual band of the electromagnetic spectrum is widely used for imaging in many applications. Infrared rays-based images are widely used, with images generated in a visible band. These types of images are used in various applications such as light microscopy, remote sensing, industry, and security. Figure 1.7 shows examples of images obtained with a visible band and an infrared band [1].

Fig. 1.8 Example of a LANDSAT satellite image

The main application of visible imaging is in remote sensing, which usually combines several bands of visual and infrared regions. This type of imaging is mainly used in NASA's LANDSAT satellite, which obtains earth images from space for weather monitoring and environmental conditions in particular areas of the Earth. An example of a LANDSAT satellite image is shown in Fig. 1.8 [3].

Also, visible imaging and infrared imaging are used in many modern industries, such as the detection of defects in product packaging, in currency, and the number recognition from vehicle number plates. Another type of imaging is microwave imaging, which is used in radars, and is taken from an unmanned aerial vehicle or airplane. Medicine and astronomy are applications of imaging in the radio band, for example, magnetic resonance imaging (Fig. 1.9) [2].

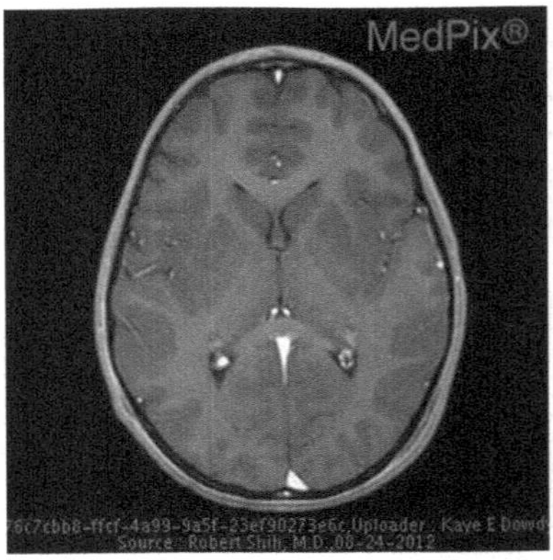

Fig. 1.9 Example of magnetic resonance imaging

There are no specific boundaries in the image, from image processing at one end to computer vision at the other. Image processing is divided into three basic types:

- *Low-level image processing:* This process involves basic operations such as noise reduction in the image, image enhancement in terms of contrast, and image sharpening. Here, the input and output of these processes are images.
- *Medium-level image processing:* This process involves operations such as image segmentation, description of the objects presented in the image, and the classification of objects. The inputs of this process are images, but the outputs are features extracted from these images, i.e., edges, contours.
- *High-level image processing:* This process involves operations such as image analysis. The inputs of this process are features of images and the outputs are also the important features of images.

The basic image processing steps involved are image acquisition, image enhancement, image restoration, color image processing, wavelets and multiresolution processing, compression, morphological processing, segmentation, representation and description, and object recognition. Image acquisition is a basic step for image processing and it is a processed image from the sense object. The inputs of image processing are images where the output of image processing may be a processed image or the attributes of an image.

In this book, theories and codes of different image processing algorithms are described for grayscale images using the Scilab software platform. Image enhancement is a simple and the most common step in image processing. It improves the quality of the image, highlighting certain features using a subjective approach. The codes and theories for image enhancement are covered in Chaps. 2 and 3. Image restoration improves the quality of images using a mathematical model of the degradation phenomenon known a priori using an objective approach. The codes and theories for image restoration are covered in Chap. 4. Morphological processing is identified as the structural information from an image. The codes and theories for morphological image processing are covered in Chap. 5. Image segmentation consists of partitioning an image into its constituents (set of pixels) to simplify and/or change the representation of an image into something that is more meaningful. The codes and theories for image segmentation are covered in Chap. 6.

1.4 Information of Scilab Software

Scilab is open-source software for the processing of mathematics equations and scientific interpretation [4, 5]. It is capable of processing various mathematics equations and automatic computing through programming. It provides all basic matrices operations, with built-in. It also provides visualization of 2D and 3D signals to help in the easy interpretation of data with which we work. Scilab is an excellent tool for beginners and teachers, in subjects that involve matrix operations. Further, there are a number of toolboxes available for developing various applications, which makes this tool very important for researchers. This software is compatible with MATLAB®, as all MATLAB M-files can be easily converted into Scilab files using the MATLAB to Scilab converter. This converter is already available in any Scilab software. The software is compatible with various operating systems, such as Windows, Linux, and MacOS X. Scilab can help a student to focus on the process for solving problems based on matrix operations.

Scilab provides an open platform for researchers and scientists who have developed various applications. Its development started around 1990 and was instigated by INRIA (the French National Institute for Research in Computer Science and Control) and ENPC (National School of Bridges and Roads) researchers [4]. Since 2003, this software has been maintained and developed by the Scilab consortium and was combined with the Digiteo Foundation in July 2008 [5]. The current version is 6.0.0 (December 2017). The early version of Scilab and its codes are freely available, but since version 5 it has been released under the GPL-compatible CeCILL license [5].

Scilab includes several mathematical functions from different languages. It has various data structures including arrays, rational functions, etc. It is a high-level programming language. Scilab has open system where any researcher can add new data types or toolboxes. Several types of toolboxes are available with Scilab [4]:

- Data analysis
- Data handling
- Documentation
- Education
- Graphical user interface
- Graphics
- Image processing
- Instruments control

- Linear algebra
- Manuals
- Modelling
- Number theory
- Numerical mathematics
- Optimization
- Physics
- Real-time
- Scilab development
- Signal processing
- Technical
- Windows tools
- XCos

Some of the basic features of Scilab are listed below [5]:

- The basic data type in Scilab is a matrix and all functions for matrix operations are built-in in Scilab.
- Graphics such as 2D and 3D can be easily generated and obtained in various formats.
- Built-in high-level programming is interpreted.

1.4.1 How to Obtain and Install Scilab

Scilab software can be downloaded directly from the Scilab homepage http://www.Scilab.org or from the download area http://www.Scilab.org/download [4]. Scilab software is available for both 32-bit and 64-bit platforms. The current version of Scilab is 6.0.0 (December 2017).

1.4.2 How to Install the Image Processing Toolbox in Scilab

In MATLAB, the whole tool box is installed when the MATLAB software is installed, but in Scilab, the user installs the tool box from the Scilab library. When starting the Scilab software, a window can be seen, as shown in Fig. 1.10. The programmer enters the Scilab commands at the prompt (**-->**). The most important menu for beginners in Scilab is the “Help” menu. By clicking on?, the “Help” menu opens with windows such as the “Help” browser with two tabs such as table of contents and search. The table of contents contains information about commands and toolboxes in alphabetical order. The “Help” browser is shown in Fig. 1.11.

Without the image processing toolbox in Scilab, codes or programs related to images cannot be performed. First, install the image processing toolbox, by typing *“atomsGui”* into the Scilab console, which gives information about the various

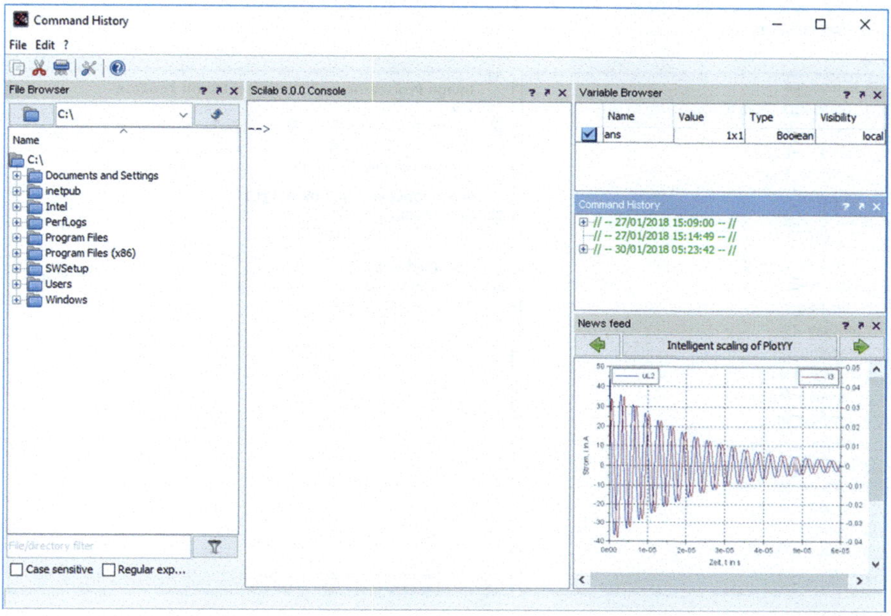

Fig. 1.10 Basic Scilab environment

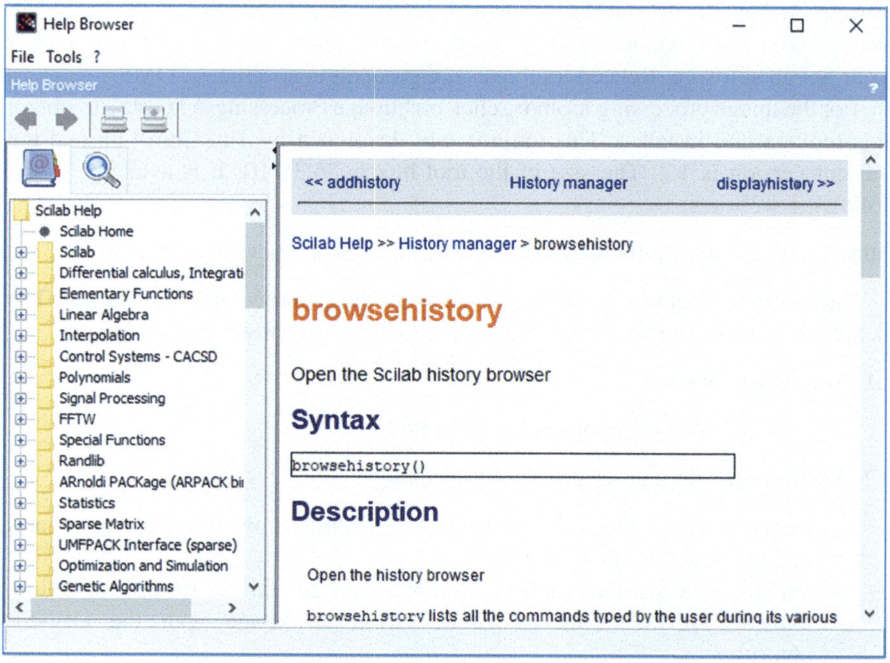

Fig. 1.11 "Help" browser of Scilab

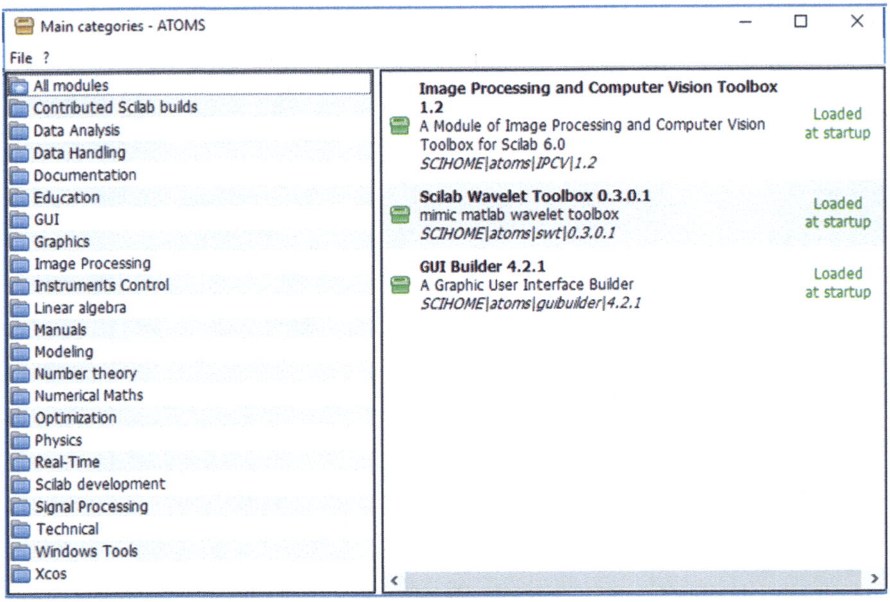

Fig. 1.12 Various AuTomatic mOdules Management for Scilab (ATOMS) in Scilab

toolboxes available for the current version of Scilab software. Figure 1.12 shows the various AuTomatic mOdules Management for Scilab (ATOMS) available for Scilab version 6.0.0. This window also gives information about toolboxes installed within your Scilab software. Related toolboxes are grouped in specific ATOMS.

For the image processing toolbox, click on "Image Processing ATOMS" to obtain this toolbox and install it. This toolbox was developed by Tan Chin Luh and the current version is 1.2. The size of the tool box is 26.3 MB. It is available at the following websites:

http://atoms.Scilab.org/toolboxes/IPCV/1.2, http://Scilabipcv.tritytech.com [6].

The toolbox contains functions for various image processing operations. The details of functions in the image processing toolbox are shown below [6]:

1. Analytic geometry
   - *plot3dot:* 3-D parametric plot for points.
2. Feature detection, description, and matching
   - *imbestmatches:* Finds the best matched features from two features objects and the matching matrix.
   - *imdetect_BRISK:* Detects features from an image with the BRISK algorithm.
   - *imdetect_DENSE:* Detects features from an image with the DENSE algorithm.

- *imdetect_FAST:* Detects features from an image with the FAST algorithm. Usually used for corner features.
- *imdetect_GFTT:* Detects features from an image with the GFTT algorithm.
- *imdetect_HARRIS:* Detects features from an image with the HARRIS algorithm.
- *imdetect_MSER:* Detects features from an image with the MSER algorithm.
- *imdetect_ORB:* Detects features from an image with the ORB algorithm.
- *imdetect_SIFT:* Detects features from an image with the SIFT algorithm.
- *imdetect_STAR:* Detects features from an image with the STAR algorithm.
- *imdetect_SURF:* Detects features from an image with the SURF algorithm.
- *imdrawmatches:* Draws a matching result for two images.
- *imextract_DescriptorBRIEF:* Computes the descriptors for a set of key points detected in an image using the BRIEF method.
- *imextract_DescriptorBRISK:* Computes the descriptors for a set of key points detected in an image using the BRISK method.
- *imextract_DescriptorFREAK:* Computes the descriptors for a set of key points detected in an image using the FREAK method.
- *imextract_DescriptorORB:* Computes the descriptors for a set of key points detected in an image using the ORB method.
- *imextract_DescriptorSIFT:* Computes the descriptors for a set of key points detected in an image using the SIFT method.
- *imextract_DescriptorSURF:* Computes the descriptors for a set of key points detected in an image using the SURF method.
- *immatch_BruteForce:* Brute-force matcher for features-matching.
- *plotFeature:* Plots the features detected by features detectors

3. Filter design, and visualization

   - *fft2pad:* Pads smaller matrix with zeros to the given size before transformation.
   - *immesh:* Visualizes a 2D matrix using a mesh plot, useful for frequency response visualization.
   - *imsmoothsurf:* Visualizes 2D matrix using a smooth surf plot, useful for frequency response visualization.
   - *imsurf:* Visualizes 2D matrix using a surf plot, useful for frequency response visualization.
   - *mkfftfilter:* Creates a frequency domain filter.

4. Image analysis and statistics

   - *corr2:* 2D correlation coefficient.
   - *edge:* Finds edges in a single channel image.
   - *imhist:* Obtains the histogram of an image.
   - *impixel:* Returns selected pixel coordinates and values.
   - *improfile:* Returns profiles for the two selected points.
   - *mean2:* Average/mean of the matrix elements.
   - *std2:* Standard deviation of the 2D matrix elements.
   - *stdev2:* Standard deviation of the 2D matrix elements.

5. Image arithmetic
   - *imabsdiff:* Calculates the absolute difference of two images.
   - *imadd:* Adds two images or adds a constant to an image.
   - *imcomplement:* Complements image.
   - *imdivide:* Divides two images or divides an image by a constant.
   - *imlincomb:* Linear combination of images.
   - *immultiply:* Multiplies two images or multiplies an image by a constant.
   - *imsubtract:* Subtracts two images or subtracts a constant from an image.
6. Image block processing
   - *im2col*: Converts the image into a series of columns.
   - *imblockproc:* Distinct block processing for an image.
   - *imblockslide:* Sliding block processing for an image.
   - *imcolproc:* Sliding block processing for an image, with vectorization.
7. Image enhancement and restoration
   - *imadjust:* Adjusts the intensity of an image from a given source histogram range to the destination histogram range.
   - *imdeconvl2:* Deconvolution with L2 regularization.
   - *imdeconvsobolev:* Deconvolution using Sobolev regularization.
   - *imdeconvwiener:* Deconvolution using the Wiener method.
   - *imdecorrstretch:* Applies decorrelation stretch to a multichannel image.
   - *imhistequal:* Histogram equalization.
   - *iminpaint:* Restores the selected region in an image using the region neighborhood.
   - immedian: Image median filter.
   - *imnoise:* Adds noise (Gaussian, etc.) to an image.
   - *imwiener2:* Wiener filter for an image.
8. Image linear filtering
   - *filter2:* 2D digital filtering.
   - *fspecial:* Creates some special 2D filters.
   - *imfilter:* Image filtering.
9. Image reading, display, and exploration
   - *imdestroy:* Destroys the graphic window created using *imdisplay* (highgui).
   - *imdestroyall:* Destroys ALL graphic windows created using *imdisplay* (*highgui*).
   - *imdisplay:* Displays image using *highgui* for a faster frame rate.
   - *imread:* Reads image file.
   - *imshow:* Displays image in a graphic window.
   - *imwrite:* Writes image to file.
   - *tifread:* Special function to read 12-bit 1024 × 1024 charge coupled device image.

10. Image registration and image fusion
    - *imfuse:* Image fusion.
    - *imgettransform:* Obtains the transformation matrix from a given source and destination points.
    - *imphasecorr:* Detection and automatic image registration for translation, rotation, and scale using the phase correlation method.
    - *imtransform:* Image affine transformation.
    - *warpmatselect:* Selecting points for image transformation.
11. Image stitching
    - *imstitchimage:* Stitches images stored in list.
    - *imstitchimage_params:* Image stitching parameters.
12. Image transforms
    - *imdct:* Discrete cosine transform (DCT).
    - *imhough:* Image Hough transformation.
    - *imhoughc:* Image Hough transformation for circle detection.
    - *imidct:* Inverse DCT.
    - *imlogpolar:* Remaps an image to log-polar space.
    - *imradon:* Calculates the 2D-Radon transform of the matrix.
13. Image types and color space conversions
    - *hsv2rgb:* Converts an hue, saturation, value (HSV) image to the equivalent red green blue (RGB) image.
    - *im2bw:* Converts image to binary.
    - *im2double:* Converts image to double precision.
    - *im2int16:* Converts image to 16-bit signed integers.
    - *im2int32:* Converts image to 32-bit signed integers.
    - *im2int8:* Converts image to 8-bit signed integers.
    - *im2uint16:* Converts image to 16-bit unsigned integers.
    - *im2uint8:* Converts image to 8-bit unsigned integers.
    - *imgraythresh:* Calculates Otsu's global threshold value.
    - *imnorm:* Normalizes input 2-D image to the range of 0–1 for double, or 0–255 for *uint8*.
    - *ind2rgb:* Converts index image to RGB image.
    - *mat2gray:* Converts matrix to grayscale image.
    - *ntsc2rgb:* Converts a National Television System Committee (NTSC) image to the equivalent RGB image.
    - *rgb2gray:* Converts RGB images to grayscale images.
    - *rgb2hsv:* Converts an RGB image to the equivalent HSV image.
    - *rgb2ind:* Converts an RGB image to an index image.
    - *rgb2lab:* Converts from RGB color space to Lab color space.
    - *rgb2ntsc:* Converts an RGB image to the equivalent NTSC image YIQ.
    - *rgb2ycbcr:* Converts an RGB image to the equivalent YCbCr image.

- *xs2im:* Converts graphics to an image matrix.
- *ycbcr2rgb:* Converts a YCbCr image to the equivalent RGB image.

14. Morphological operations
    - *bwborder:* Finds a border for an image.
    - *imblackhat:* Image *blackhat.*
    - *imblobprop:* Calculates blobs properties from a labeled image.
    - *imclose:* Image closing.
    - *imcreatese:* Creating structuring elements for morphological operation.
    - *imdilate:* Image dilation.
    - *imerode:* Image erosion.
    - *imfill:* Filling holes for objects in a binary image.
    - *imgradient:* Image gradient.
    - *imhitmiss:* Image hit–miss.
    - *imlabel:* Finds blobs in an image.
    - *imopen:* Image opening.
    - *imtophat:* Image *tophat.*
15. Object detection
    - *imdetectobjects:* Detects objects in an image with cascade classification.
16. ROI processing
    - *imroifill:* Fills an image using the border color of the selected region.
    - *imroifilt:* Filtering of a selected region.
17. Spatial transformations
    - *imcrop:* Crops an image.
    - *imcropm:* Crops an image using mouse selection.
    - *impyramid:* Image pyramid reduction and expansion.
    - *imresize:* Resizes image.
    - *imrotate:* Rotates an image to a given angle.
18. Structural analysis and shape descriptors
    - *imconvexhull:* Finds the convex hull of a point set.
    - *imdrawcontours:* Draws contours from the contour image.
    - *imfindcontours:* Finds contours in a binary image.
19. Super resolution
    - *imsuperres:* Super resolution with image sequences.
    - *imsuperres_params:* Super resolution with image sequences parameters.
20. Utilities and interactive tools
    - *im2movie:* Creates a movie from a sequence of images.
    - *imaddtext:* Adds text to a color image.
    - *imbreakset:* Sets the break event with Scilab figure.

- *imbreakunset:* Unsets the break event with Scilab figure.
- *imcaminfo:* Shows the supported raw resolution for a USB camera (Linux only).
- *imdistline:* Measures the distance between two selected points in pixels.
- *imlsusb:* Lists all USB devices connected to a PC (Linux only).
- *impixelval:* Interactive tool to inspect pixel value at a selected point.
- *imrects:* Draws bounding boxes on an image.
- *imroi:* Selects region of interest and creates a mask from it.
- *imselect:* Selects points on an image.
- *rectangle:* Draws a rectangle on an image.

21. Video handling
    - *addframe:* Adds a frame to the video file.
    - *aviclose:* Closes a video file.
    - *avicloseall:* Closes all opened video files/cameras.
    - *avifile:* Creates a new video file to write.
    - *avilistopened:* Shows all open video files.
    - *aviopen:* Opens a video file.
    - *avireadframe:* Grabs and returns a frame from an opened video file or camera.
    - *camopen:* Opens a camera.
    - *camread:* Grabs and returns a frame from a camera.

1.5 Areas of Image Processing Covered in the Book

This chapter briefly discussed typical characteristics of signals, images, and basic image processing. The rest of this book is organized as follows.

Chapter 2: Image Enhancement in the Spatial Domain This chapter covers in detail how to use Scilab for image enhancement in the spatial domain and spatial filters.

Chapter 3: Image Enhancement in the Frequency Domain This chapter covers in detail how to use Scilab for image enhancement in the frequency domain and domain filters.

Chapter 4: Image Restoration This chapter covers in detail how to use Scilab for image restoration and various filters.

Chapter 5: Morphological Image Processing This chapter covers in detail how to use Scilab for morphological image processing, various morphological theories, and their implementation in Scilab.

Chapter 6: Image Segmentation This chapter explains the various functions of Scilab for image segmentation.

Chapter 7: Color Image Processing This chapter covers in detail how to use Scilab for color image processing.

Bibliography

1. Gonzalez, R., & Woods, R. (2008). *Digital image processing*. Upper Saddle River: Pearson Education India.
2. MedPix™ medical image database available at http://rad.usuhs.mil/medpix/medpix.html, https://medpix.nlm.nih.gov/home.
3. University of South Carolina SIPI image database. http://sipi.usc.edu/database/database.php.
4. Scilab software. http://www.scilab.org/download. Last accessed November 2017.
5. Annigeri, S. (2009). *An introduction to Scilab*. Hubli: B. V. Bhoomaraddi College of Engineering and Technology.
6. Image processing toolbox of Scilab. http://atoms.scilab.org/toolboxes/IPCV/1.2, http://scilabipcv.tritytech.com. Last accessed November 2017.

Chapter 2
Image Enhancement in the Spatial Domain

2.1 Introduction

Whenever an image is represented in terms of a collection of pixels, it is said to be in the spatial domain. A method that directly operates on the pixel values to process an image is called a spatial domain method. All spatial domain methods can be represented using the following general equation:

$$g(x, y) = P(f(x, y)) \tag{2.1}$$

where $f(x, y)$ is an input image, $g(x, y)$ is a processed image, and P is a processing operator.

A neighborhood pixel (x, y) is generally defined as a portion of the sub-image which, when centered, lies at (x, y). The center of the sub-image is moved from point to point and the processing operation P is applied at each point position to generate a processed point at that position. This processing is called image processing in the spatial domain and is depicted in Fig. 2.1. This processing is divided in two types: point processing and neighborhood processing [1, 2].

2.2 Image Enhancement by Point Processing

Image enhancement by point processing is the simplest form of image processing in the spatial domain. It is also called gray level transformation. Here, the size of P is 1×1 and it means that the value of $g(x, y)$ depends only on the original value of $f(x, y)$. In this case, P becomes a gray level transformation function of the following form:

$$S = P(r) \tag{2.2}$$

R. M. Thanki, A. M. Kothari, *Digital Image Processing using SCILAB*,
https://doi.org/10.1007/978-3-319-89533-8_2

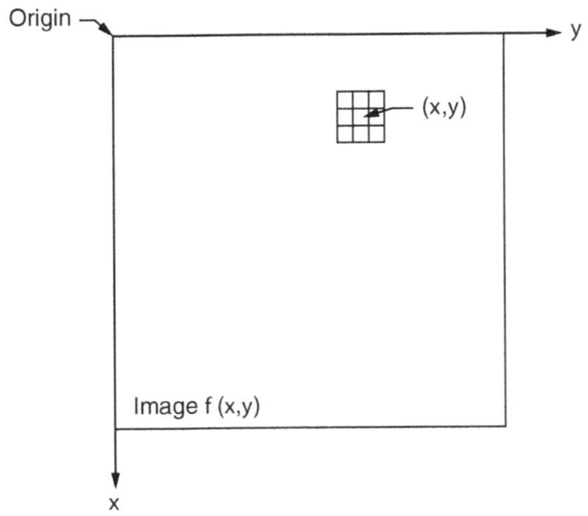

Fig. 2.1 Image processing in the spatial domain

where r is the gray level of the original image $f(x, y)$ and S is the gray level of the processed image $g(x, y)$.

Some of the important point processing methods for images are as follows:

- Identity transformation
- Image negative
- Contrast stretching
- Contrast thresholding
- Gray level slicing
- Bit plane slicing
- Log transformation
- Power law transformation.

2.2.1 Identity Transformation

In identity transformation, the processed pixel will have the same gray level as the original pixel. Mathematically, this transformation is given as follows:

$$S = r \tag{2.3}$$

where $0 \le r \le L - 1$, $0 \le S \le L - 1$, L being the total number of gray levels.

If the image is an 8-bit image then the gray level L value $= 256$, and therefore $0 \le r \le 255$, $0 \le S \le 255$. Graphically, this transformation is shown in Fig. 2.2.

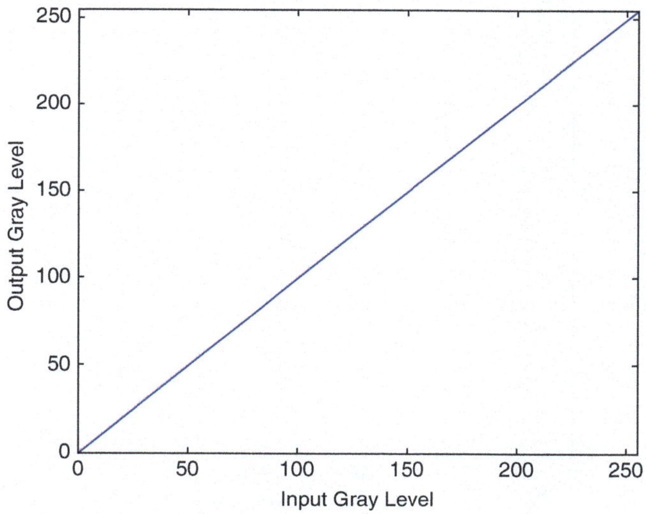

Fig. 2.2 Identity transformation

2.2.2 Image Negative

Image negative transformation means inversion of the gray level, i.e., the black value pixel in the original image will become the white value pixel in the processed image and vice versa. Mathematically, this transformation is given follows:

$$S = L - r - 1 \tag{2.4}$$

where $0 \leq r \leq L - 1$, $0 \leq S \leq L - 1$, L being the total number of gray levels.

If the image is an 8-bit image, then the gray level L value $= 256$, and therefore $0 \leq r \leq 255$, $0 \leq S \leq 255$. Graphically, this transformation is shown in Fig. 2.3.

The SCILAB code for this transformation is given here and the output of this code is shown in Fig. 2.4.

SCILAB Code for Image Negative Transformation

```
a=uigetfile('*.*','Select the Image:-');
a=imread(a);
new=255 - a;
new=double(new);
imshow(new);
```

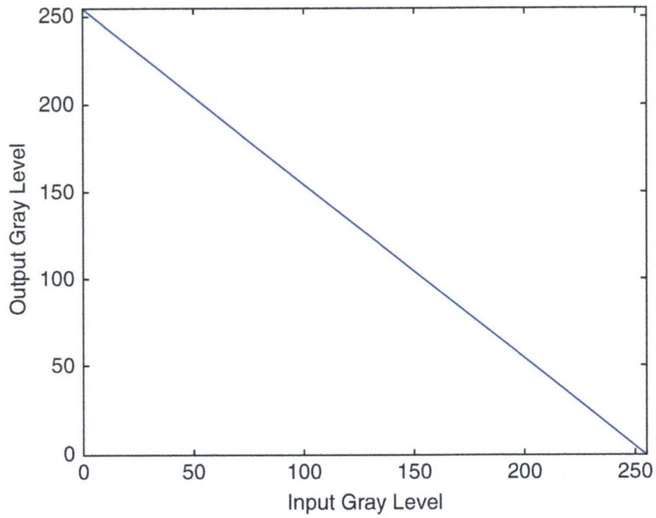

Fig. 2.3 Image negative transformation

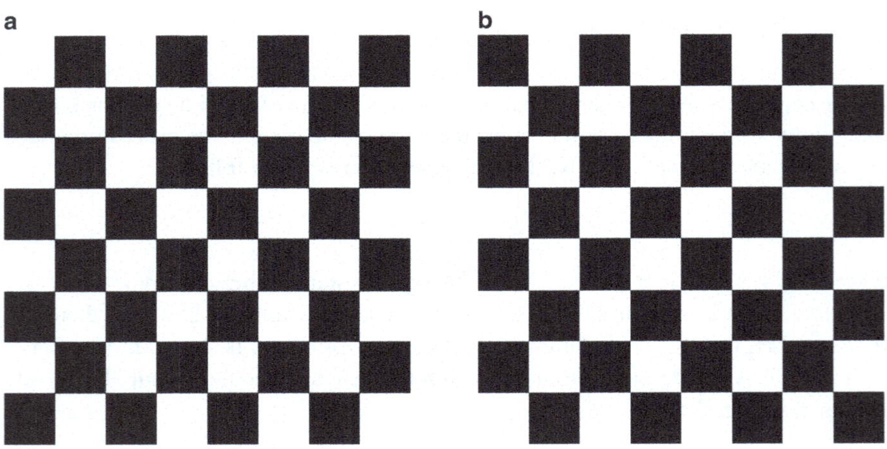

Fig. 2.4 Result of image negative transformation: (**a**) original image; and (**b**) processed image

2.2.3 Contrast Stretching

The contrast stretching transformation is used to improve details of images that are acquired using poor illumination conditions, an image sensor with a less dynamic range, or the wrong camera lens setting. The output of this transformation is an image with higher contrast than the original image, which is achieved by darkening

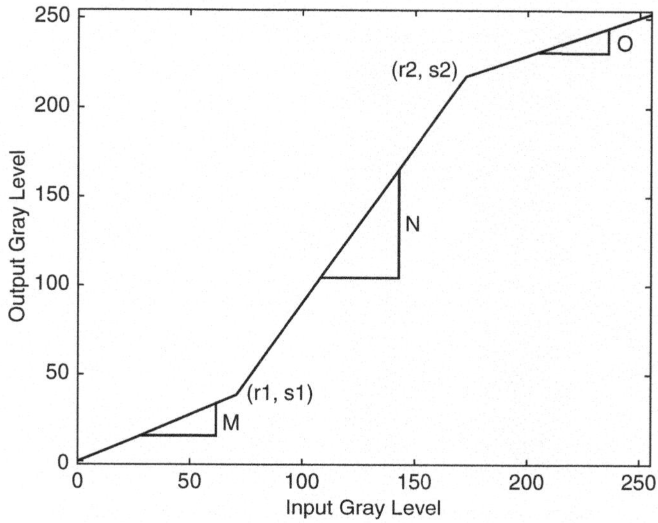

Fig. 2.5 Contrast stretching

levels below gray level $r1$ and brightening levels above $r2$ in the original image. Mathematically, this transformation is given as follows:

$$S = Mr, 0 \leq r \leq r1 \tag{2.5}$$

$$S = N(r - r1) + s1, r1 \leq r \leq r2 \tag{2.6}$$

$$S = O(r - r1) + s2, r2 \leq r \leq L - 1 \tag{2.7}$$

Based on Eqs. 2.5 and 2.6, this transformation is divided into two special cases: linear transformation if $r1 = r2$ and $s1 = s2$ and contrast thresholding if $r1 = r2$, $s1 = 0$, and $s2 = L - 1$. Graphically, this transformation is shown in Fig. 2.5.

The SCILAB code for this image processing is given here and the output of this code is shown in Fig. 2.6.

SCILAB Code for Contrast Stretching

```
a=uigetfile('*.*','Select the Image:-');
a=imread(a);
a1=min(min(a));
b=max(max(a));
l=0.5;
n=0.5;
m=3;
v=l*a1;
w=v+(m*(b-a1));
```

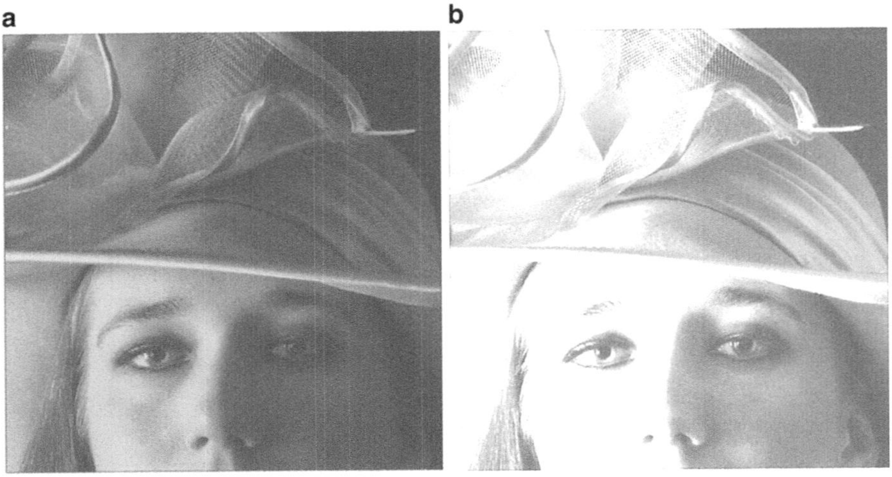

Fig. 2.6 Result of contrast stretching: (**a**) original image; and (**b**) processed image

```
[r,c]=size(a);
for i=1:r
  for j=1:c

    if (a(i,j)<a1)
      new(i,j)=l*a(i,j);
    elseif (a(i,j)>=a1 & a(i,j)<b)
      new(i,j)=(m*(a(i,j)-a1))+v;
    else
      new(i,j)=(n*(a(i,j)-b))+w;
    end
  end
end
new=double(new);
imshow(new);
```

2.2.4 Contrast Thresholding

Another type of transformation is the special case of contrast stretching where $r1 = r2$, $s1 = 0$, and $s2 =$ L – 1. Mathematically, this transformation is given as follows:

$$S = 0, 0 \leq r \leq r1 \tag{2.8}$$

$$S = L - 1, r1 \leq r \leq L - 1 \tag{2.9}$$

Fig. 2.7 Result of contrast thresholding: (a) original image; (b) processed image

If the original image is an 8-bit image, then $L = 256$ and, therefore, $S = 255$, $rl \leq r \leq 255$. The SCILAB code for this image processing is given here and the output of this code is shown in Fig. 2.7.

SCILAB Code for Contrast Thresholding

```
a=uigetfile('*.*','Select the Image:-');
a=imread(a);
T=58; //threshold value
 [r,c]=size(a);
 for i=1:r
   for j=1:c
     if (a(i,j)<=T)
       x(i,j)=0;
     else
       x(i,j)=255;
     end
   end
end
x=uint8(x);
imshow(x);
```

2.2.5 *Gray Level Slicing*

Gray level slicing is used to highlight gray levels in the range of interest. This approach takes two different forms: gray level slicing without background and gray level slicing with background.

2.2.5.1 Gray Level Slicing Without Background

In the gray level slicing without background approach, a high value is assigned to the pixels with gray levels in the range of interest and a low value is assigned to the other pixels. Mathematically, this transformation is given as follows:

$$S = L - 1, A \leq r \leq B; \;\; S = 0, \;\; \text{otherwise} \tag{2.10}$$

If the original image is an 8-bit image, then $L = 256$ and, therefore, $S = 255$, $A \leq r \leq B$. Graphically, this transformation can be seen in Fig. 2.8a.

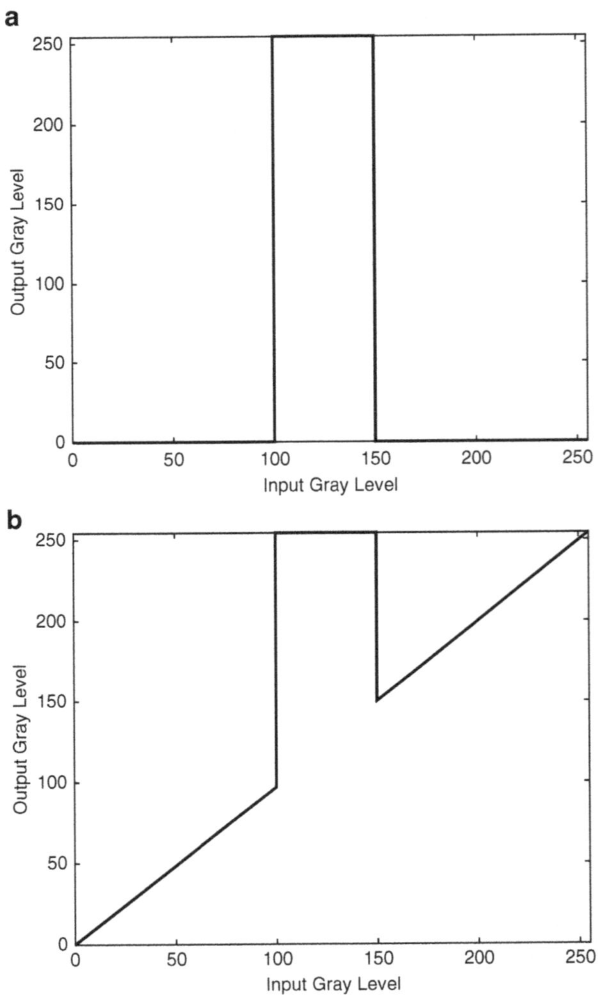

Fig. 2.8 Gray level slicing (**a**) without background; and (**b**) with background

Fig. 2.9 Result of gray level slicing (**a**) without background; and (**b**) with background

2.2.5.2 Gray Level Slicing with Background

The problem with the gray level slicing without background approach is that the entire background is destroyed. To preserve the background and to brighten the desired range, pixels in the desired range are assigned a high value while the other pixels are not altered. Mathematically, this transformation is given as follows:

$$S = L - 1, A \leq r \leq B; \ \ S = r, \ \text{otherwise} \tag{2.11}$$

If the original image is an 8-bit image, then $L = 256$ and, therefore, $S = 255$, $A \leq r \leq B$. Graphically, this transformation can be seen in Fig. 2.8b.

The SCILAB codes for gray level slicing without and with background are given here and the output of these codes is shown in Fig. 2.9.

SCILAB Code for Gray Level Slicing Without Background

```
a=uigetfile('*.*','Select the Image:-');
a=imread(a);
a1=58; // This value is user defined
b1=158; // This value is user defined
[r,c]=size(a);
for i=1:r
  for j=1:c


    if (a(i,j)>a1 & a(i,j)<b1)
      x(i,j)=255;
    else
      x(i,j)=0;
    end
  end
end
x=uint8(x);
imshow(x);
```

SCILAB Code for Gray Level Slicing With Background

```
a=uigetfile('*.*','Select the Image:-');
a=imread(a);
a1=170; // This value is user defined
b1=190; // This value is user defined
[r,c]=size(a);
for i=1:r
  for j=1:c
    m=a(i,j);
    if ((m>a1) & (m<b1))
      x(i,j)=255;
    else
      x(i,j)=a(i,j);
    end
  end
end
x=uint8(x);
imshow(x);
```

2.2.6 *Bit Plane Slicing*

A gray-scale image can be represented using a sequence of binary images by breaking it into different bit-planes. If the gray value of each pixel of the image is represented by 8-bit, then the last bit has the lowest value and is called the least significant bit. The seventh bit has the highest value and is called the most significant bit. This type of image processing is mainly used in image compression and image watermarking. The bit plane is reconstructed in a reverse process of its slicing to generate the processed image. For image compression, some data relating to the bit plane can be ignored which is less significant for representation of the image. Figure 2.10 provides a basic representation of bit plance slicing.

The SCILAB code for bit plane slicing is given here and the output of this code is shown in Fig. 2.11.

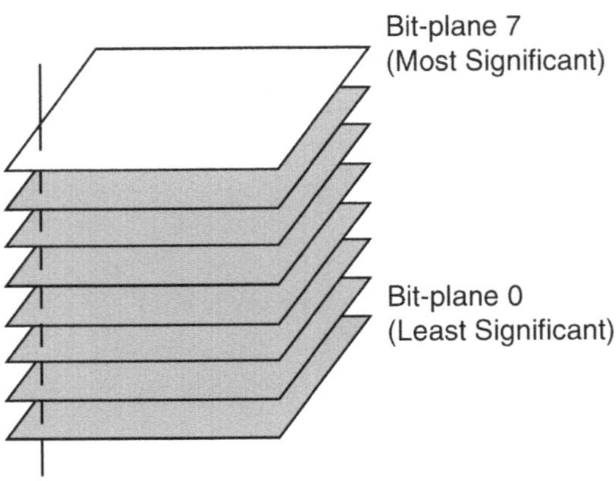

Fig. 2.10 Basic idea of bit plane slicing

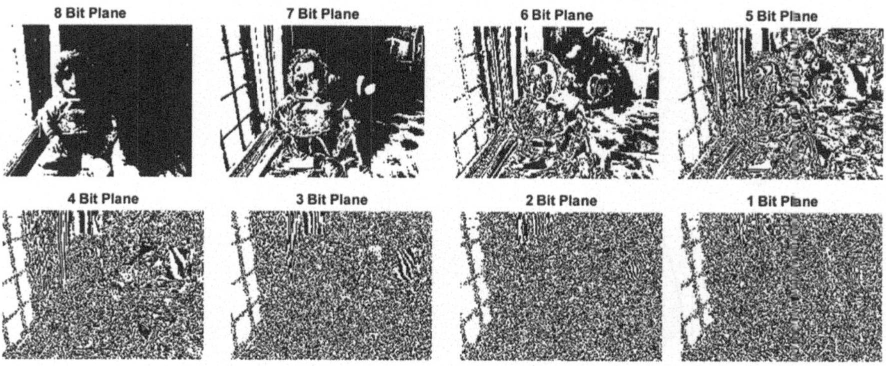

Fig. 2.11 Result of bit plane slicing

SCILAB Code for Bit Plane Slicing

```
f=uigetfile('*.*','Select the Image:-');
f=imread(f);
f=double(f);
[r,c]=size(f);
com=[128 64 32 16 8 4 2 1];
for k=1:1:length(com);
  for i=1:r
    for j=1:c
        new(i,j)=bitand(f(i,j),com(k));
    end
  end
end
```

2.2.7 Log Transformation

During daytime, we can't observe the stars in the sky due to the presence of the sun and its high intensity. In the same way, it is suggested that our eye cannot see images with high pixel values. Thus, some pixel values must lie in a certain meaningful range in order to be obscured. Log transformation is used to compress the high pixel values so that the user can see the image clearly. The log operator is an excellent compressing function and is therefore used in this transformation. It expands the values of dark pixels and compresses the high pixel values. Mathematically, this transformation is given as follows:

$$S = c \times \log(1 + r) \tag{2.12}$$

where c is a constant value

Graphically, this transformation can be viewed in Fig. 2.12.

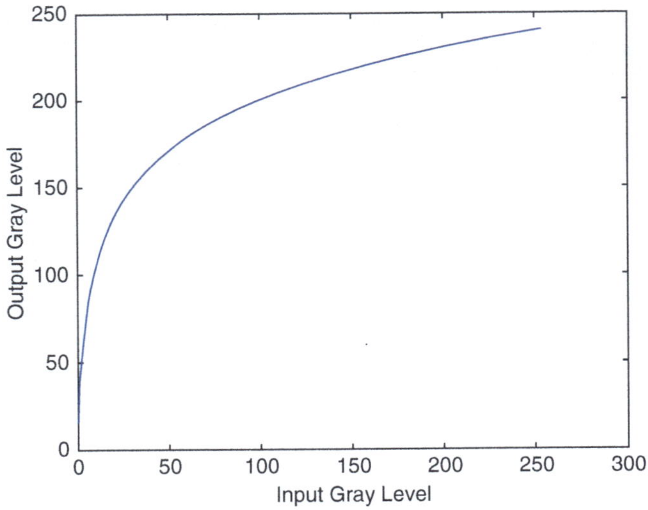

Fig. 2.12 Log transformation

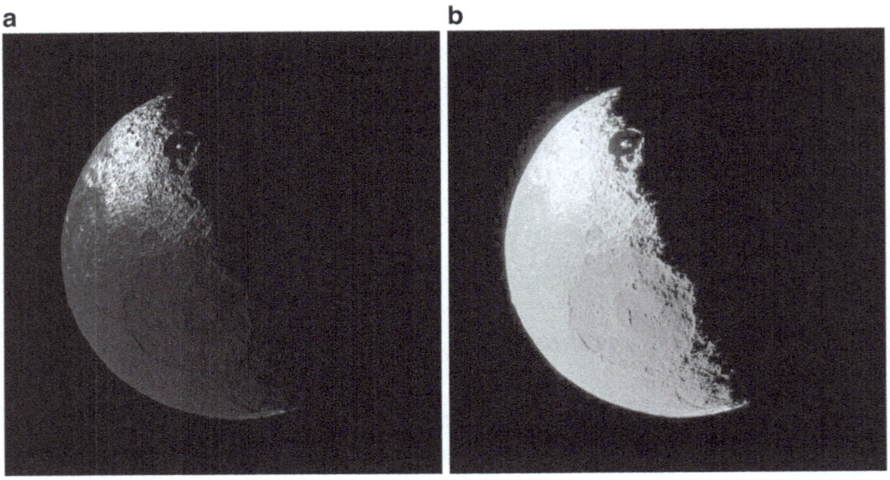

Fig. 2.13 Result of log transformation: (**a**) original image; (**b**) processed image

The SCILAB code for log transformation is given here and the output of this code is shown in Fig. 2.13.

SCILAB Code for Log Transformation

```
a=uigetfile('*.*','Select the Image:-');
a=imread(a);
c=1;
```

```
[r1,c1]=size(a);
for i=1:r1
  for j=1:c1
    b=double(a(i,j));
    s(i,j)=c*log10(1+b);
  end
end
new=s;
new1=uint8(new*100);
imshow(new1);
```

2.2.8 Power Low Transformation

The power low transformation maps a narrow range of dark input values into a wider range of output values or vice versa. Varying γ gives a whole family of curves. Mathematically, this transformation is given as follows:

$$S = c \times r^{\gamma} \tag{2.13}$$

where c is a constant value

Graphically, this transformation can be viewed in Fig. 2.14. This method is widely used in television for the correction of video representations using cathode ray tubes (CRTs).

The SCILAB code for power low transformation is given here and the output of this code is shown in Fig. 2.15.

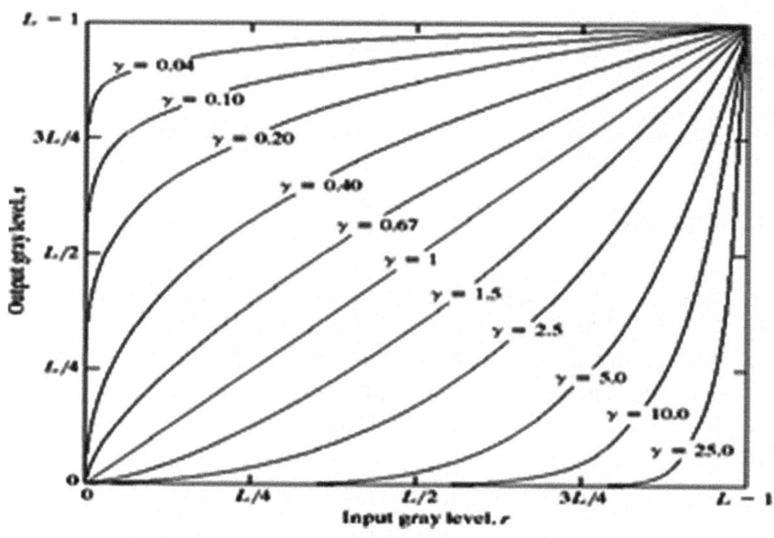

Fig. 2.14 Power low transformation

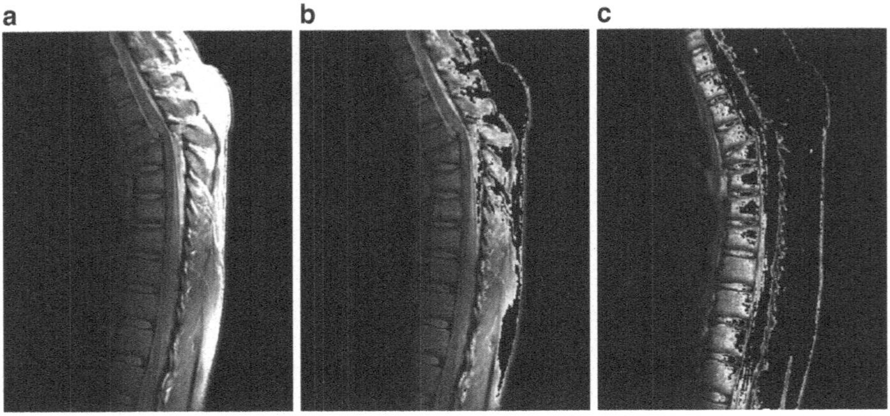

Fig. 2.15 Result of power low transformation: (**a**) original image; (**b**) processed image with $\gamma = 0.8$; and (**c**) processed image with $\gamma = 1.2$

SCILAB Code for Power Low Transformation

```
a=uigetfile('*.*','Select the Image:-');
a=imread(a);
r=size(a,1);
c=size(a,2);
G=0.6;
//G=input('Enter the value of Gamma:-');
for i=1:r
  for j=1:c
    b=double(a(i,j));
    x(i,j)=b.^G;
  end
end
new=x;
new1=uint8(new);
imshow(new1);
```

2.3 Histogram

The histogram of an image represents the relative frequency of occurrence of the various gray levels in the image. It provides a total description of the appearance of an image. The type and degree of enhancement obtained depends on the nature of the specified histogram. The histogram of a digital image with gray levels in the range $[0, L - 1]$ is a discrete function given as follows:

$$h(r_k) = n_k \tag{2.14}$$

where $k = 0, \ldots, L - 1$, r_k is the kth gray level, and n_k is the number of pixels with gray level r_k. As an example, a simple matrix and its histogram is shown in Fig. 2.16.

a

2	1	2	3
2	1	3	3
4	3	4	3
4	2	4	3

b

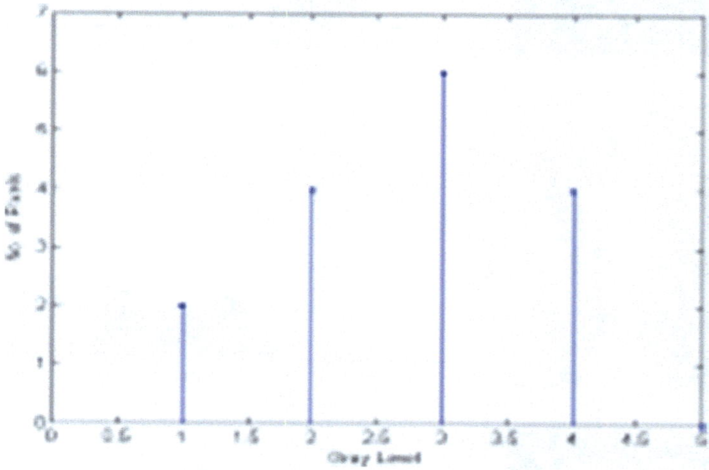

Fig. 2.16 Simple matrix and its histogram

By looking at the histogram, we can identify the type of image, as shown in the examples in Fig. 2.17.

In a dark image, the components of the histogram are concentrated on the low side of the gray levels. For bright images, histogram components are concentrated near the high side of the gray scale. For low-contrast images, the histogram is narrow and centered near the middle of the gray scale. For high-contrast images, the histogram covers a broad range of the gray scale and distribution of pixels is almost uniform with very few high spikes.

The SCILAB code for the histogram without using the inbuilt *imhist* function is given here and the output of this code is shown in Fig. 2.18.

SCILAB Code for Histogram

```
a=uigetfile('*.*','Select the Image:-');
a=imread(a);
r=size(a,1);
c=size(a,2);
h=zeros(1,256);
for i=1:r
```

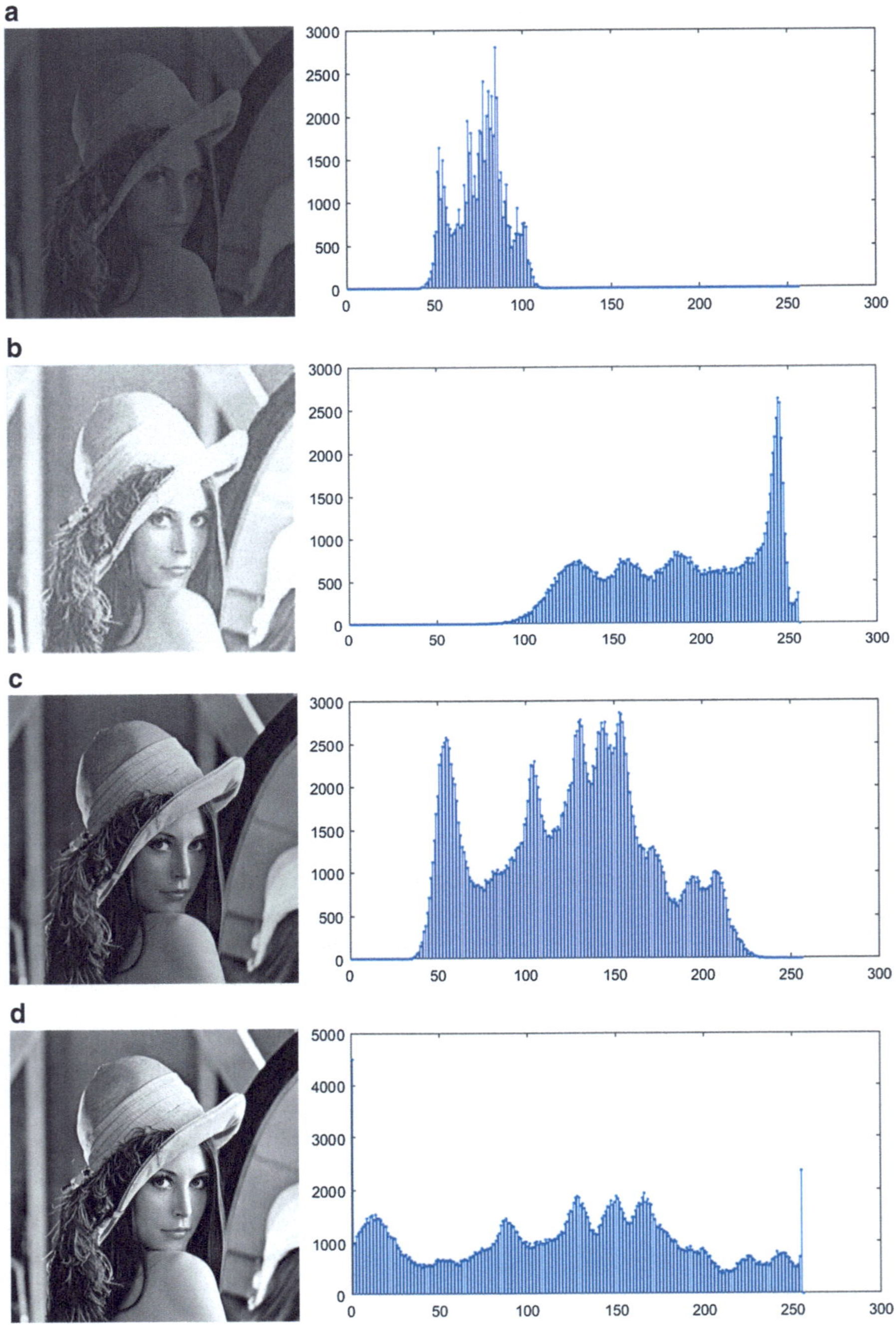

Fig. 2.17 Four basic types of images and their histogram: (**a**) dark image; (**b**) bright image; (**c**) low-contrast image; and (**d**) high-contrast image

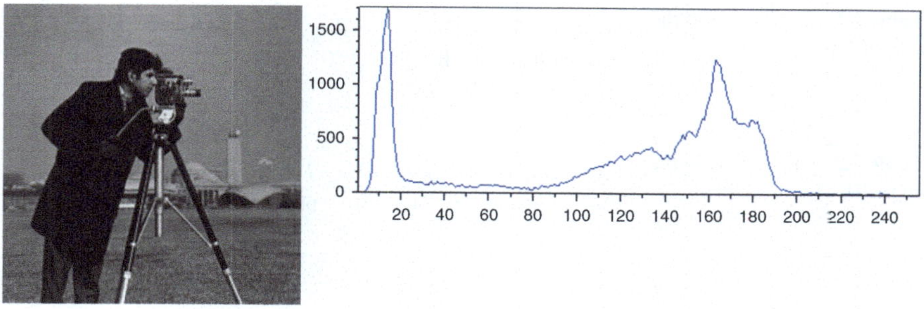

Fig. 2.18 Result of an image histogram

```
    for j=1:c
      if (a(i,j)==0)
         a(i,j)=1;
      end
      k=a(i,j);
      h(k)=h(k)+1;

    end
  end
  plot(h);
```

2.3.1 *Histogram Processing*

Histogram processing can be used for image enhancement, image compression, and image segmentation and can be divided into two types: histogram stretching and histogram equalization. Details of these two forms of histogram processing are given in Sects. 2.3.1.1 and 2.3.1.2.

2.3.1.1 Histogram Stretching

The process of histogram stretching increases the dynamic range of the image and, hence, improves the contrast of the image. In this process, the basic shape of the histogram is not modified but the entire range of histogram values is stretched. Before the stretching process, it is necessary to specify the upper and lower limits for pixels of the normalized image. These limit values depend on the type of image. For example, for 8-bit images, the minimum limit value is 0 and the maximum limit value is 255. In Eq. 2.15, the simplest normalization of the image to find the lowest and highest pixel value of an image are represented by c and d, respectively. The minimum and maximum limit values are denoted with a and b, respectively. The histogram stretching function F for pixels is given as follows:

Table 2.1 Gray levels in the range [2, 6]

Gray level	0	1	2	3	4	5	6	7
No. of pixels	0	0	50	60	50	20	10	0

Table 2.2 Gray levels in the range [0, 7]

New gray level	0	1	2	3	4	5	6	7
No. of pixels	50	0	60	0	50	20	0	10

$$F = (r - c) \times \frac{(b - a)}{(d - c)} + a \tag{2.15}$$

Example of Histogram Stretching

Table 2.1 shows the gray levels in the range [2, 6].

Histogram stretching should be performed so that the new image has the dynamic range of [0, 7].

Solution

Here, $a = 0$, $b = 7$, $c = 2$, and $d = 6$. After performing the operation using Eq. 2.15 we can achieve the following table (Table 2.2).

The SCILAB code for histogram stretching is given here and the output of this code is shown in Fig. 2.19.

SCILAB Code for Histogram Stretching

```
a=uigetfile('*.*','Select the Image:-');
a=imread(a);
r=size(a,1);
c=size(a,2);
rmax=(max(a));
rmin=(min(a));
smin=0;
smax=255;

constant = (smax-smin)/(rmax-rmin);
for i=1:r
  for j=1:c
    x(i,j)=(constant*(a(i,j)-rmin))+smin;
  end
end
h1=imhist(x);
figure,imshow(x);
subplot(2,1,2);plot(h1);
```

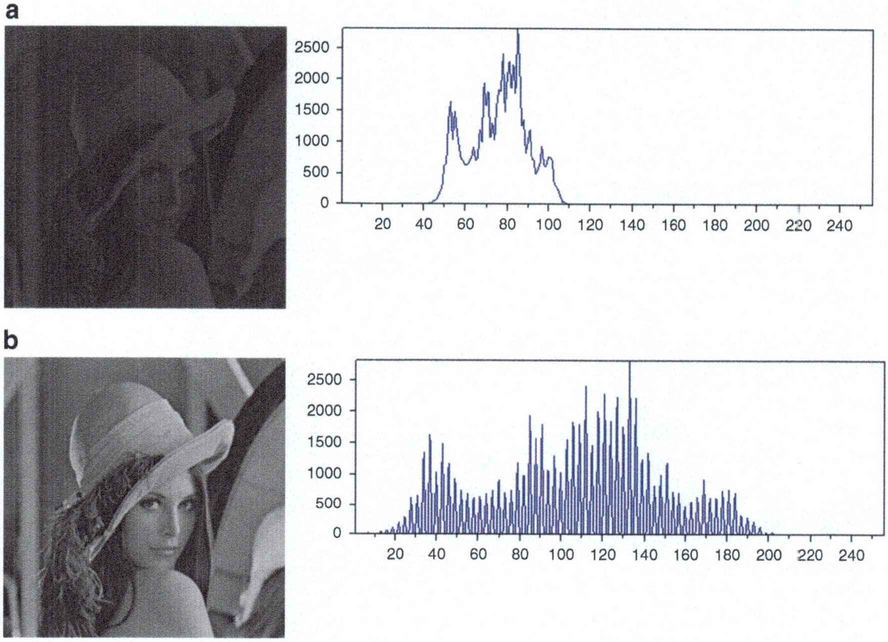

Fig. 2.19 Result of histogram stretching: (**a**) original image and its histogram; and (**b**) processed image and its histogram

2.3.1.2 Histogram Equalization

In histogram equalization, the main idea is to equalize all values of the histogram so that the entire gray level range from 0 to $L - 1$ is covered. Transformation for this process may be as follows:

$$S = T(r), 0 \leq r \leq 1 \tag{2.16}$$

The following conditions can be used to satisfy Eq. 2.16:

- $T(r)$ is a single-valued and monotonically increasing function in the interval $0 \leq r \leq 1$.
- $0 \leq T(r) \leq 1$ for $0 \leq r \leq 1$.

The first condition is needed to guarantee that the inverse transformation will exist and monotonicity ensures that no gray level is inverted. The second condition ensures that the output gray levels will be in the same range as the input levels.

The inverse transformation from S back to r is given as follows:

$$r = T^{-1}(S), 0 \leq S \leq 1 \tag{2.17}$$

The gray levels in the image can be viewed as random variables in the interval [0, 1]. They can be characterized by their probability density function (PDF). From probability theory, if $P_r(r)$ is known and $T(r)$ satisfies the above conditions, then the PDF of output gray levels $P_S(S)$ can be given as follows:

$$P_S(S) = P_r(r)\left|\frac{dr}{ds}\right| \tag{2.18}$$

Consider the following transformation function:

$$S = T(r) = \int_0^r P_r(w)dw \tag{2.19}$$

This equation is called the cumulative distribution function (CDF) of input gray levels r. Taking the differentiation of Eq. 2.19 with reference to r gives Eq. 2.20:

$$\frac{dS}{dr} = \frac{d}{dr}T(r) = \frac{d}{dr}\left[\int_0^r P_r(w)dw\right] = P_r(r) \tag{2.20}$$

Substituting this result in Eq. 2.18 gives Eq. 2.21:

$$P_s(S) = P_r(r)\frac{1}{P_{r(r)}} = 1 \tag{2.21}$$

Equation 2.21 is called the uniform probability density function. So, taking the CDF of r as the transformation function produces an image with gray levels that have uniform probability. For digital image processing (DIP), we have to formulate the above equations in discrete form. The discrete form of Eq. 2.19 is as follows:

$$S_k = T(r_k) = \sum\nolimits_{i=0}^{k} P_r(r_j) = \sum\nolimits_{i=0}^{k} \frac{n_i}{n}, \text{ for } k = 0, 1, \ldots, L-1 \tag{2.22}$$

Equation 2.22 is called histogram equalization.

Example of Histogram Equalization

Table 2.3 shows the numbers of pixels for an image with gray levels in the range [0, 7].

Solution for Histogram Equalization

A solution for histogram equalization is shown in Tables 2.4 and 2.5.

Table 2.3 Gray levels in the range [0, 7]

Gray level	0	1	2	3	4	5	6	7
No. of pixels	790	1023	850	656	329	245	122	81

Table 2.4 Histogram equalization process on given gray level values

Gray level	n_k	PDF	CDF.	$(L - 1) \times S_k$	Rounding off
0	790	0.19	0.19	1.33	1
1	1023	0.25	0.44	3.08	3
2	850	0.21	0.65	4.55	5
3	656	0.16	0.81	5.67	6
4	329	0.08	0.89	6.23	6
5	245	0.06	0.95	6.65	7
6	122	0.03	0.98	6.86	7
7	81	0.02	1	7	7
Total	4096				

PDF, Probability Density Function
CDF, Cummulative Density Function

Table 2.5 Output of histrogram equalization process

Equalized gray level	0	1	2	3	4	5	6	7
No. of pixels	0	790	0	1023	0	850	985	448

Note Why not a flat histogram?

The discrete domain is an approximation of the continuous domain, i.e., values between 0 and 1, 1 and 2, and so on, are not known. Hence, we cannot achieve a perfectly flat histogram.

The SCILAB code for histogram equalization is given here and the output of this code is shown in Fig. 2.20.

SCILAB Code for Histogram Equalization

```
a=uigetfile('*.*','Select the Image:-');
a=imread(a);
a=double(a);
big=max(max(a));
[r c]=size(a);
tot = r*c;
h=zeros(1,256); // to store the histogram values
z=zeros(1,256);
for i=1:1:r
  for j=1:1:c
    if a(i,j)==0
      a(i,j)=1;
    end
  end
end
for i=1:1:r
  for j=1:1:c
    t=a(i,j);
    h(t)=h(t)+1;
  end
end
pdf=h/tot;
cdf(1)=pdf(1);
```

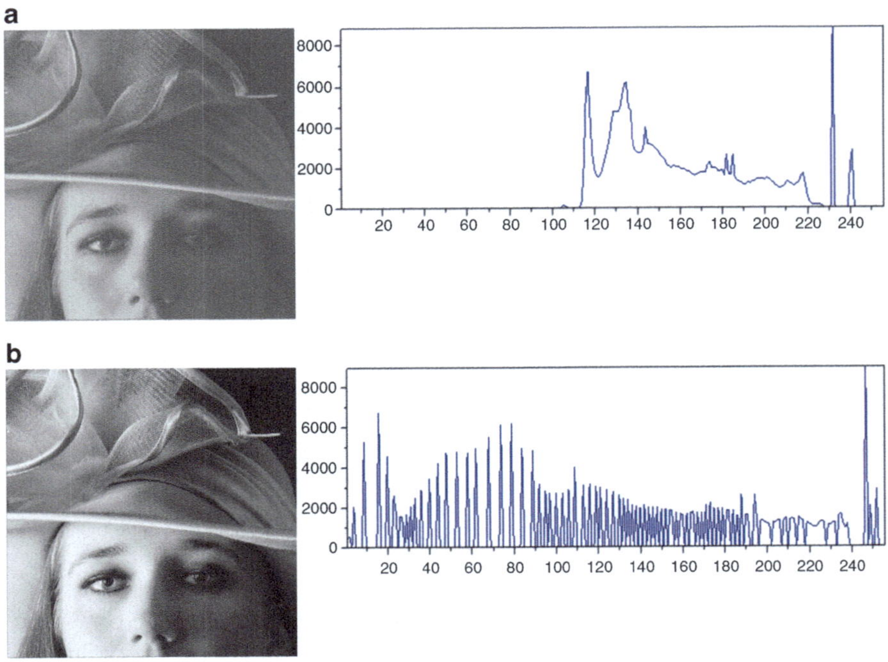

Fig. 2.20 Result of histogram equalization: (**a**) original image and its histogram; and (**b**) processed image and its histogram

```
for i=2:1:big
  cdf(i)=pdf(i)+cdf(i-1);
end
new=round(cdf*big);
new=new+1;
for i=1:1:r
  for j=1:1:c
    temp=a(i,j);
    b(i,j)=new(temp);
    t=b(i,j);
    z(t)=z(t)+1;
  end
end
subplot(2,1,2);plot(z);
```

2.3.1.3 Histogram Matching (Specification)

Histogram matching is a process where a time series, image, or higher dimension scalar data is modified such that its histogram matches that of another (reference) dataset. A common application of this is to match the images from two sensors with slightly different responses, or images from a sensor whose response changes over time.

2.4 Image Enhancement by Neighborhood Processing: Spatial Domain Filters

We have seen in Sect. 2.2 that an image can be enhanced by applying various gray level transformations to each value of the pixel. Neighbor processing is an extension process of gray level transformation, where an operating function is applied to a neighborhood pixel of each targeted pixel. This processing is based on a *mask* process. In this process, a new image will be created whose pixel values have gray level values calculated from the gray level values under the mask. This process is shown in Fig. 2.21.

The combination of the mask and operating function is called the image filter. If this filter gives a new gray level value using a linear operation, it is called a linear filter. This filter can be implemented by multiplying all values in the mask by corresponding values in the neighborhood pixels and adding all values together. Suppose a 3 × 5 mask as shown in Fig. 2.22. The values of this mask are given as shown in Table 2.6 and the corresponding values of pixels are given as in Table 2.7.

The processed pixel value can be gained using the following equation:

$$p(x,y) = \sum_{s=-1}^{1} \sum_{t=-2}^{2} M(s,t)P(x+s, y+t) \tag{2.23}$$

This process is known as spatial filtering and is shown in Fig. 2.22.

For spatial filtering, three basic steps are required for image enhancement:

- Position the mask over the current pixel.
- Form all products of filter elements with the corresponding elements of the neighborhood.
- Add up all the products.

This process is repeated for every pixel in the image.

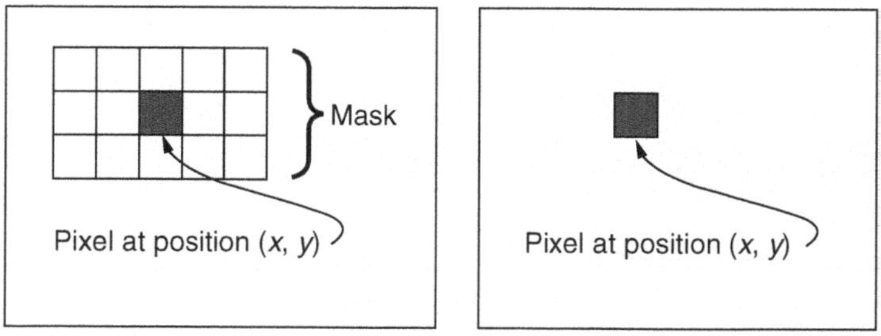

Fig. 2.21 Mask processing: (**a**) original image; and (**b**) image after filtering

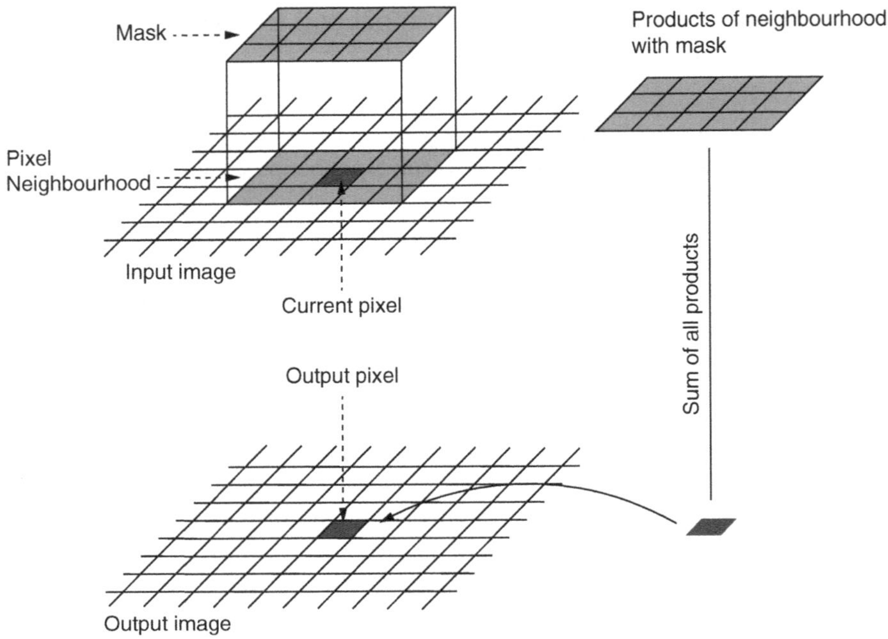

Fig. 2.22 Spatial filtering process

Table 2.6 Values of the mask used in the image filter

$F(-1, -2)$	$F(-1, -1)$	$F(-1, 0)$	$F(-1, 1)$	$F(-1, 2)$
$F(0, -2)$	$F(0, -1)$	$F(0, 0)$	$F(0, 1)$	$F(0, 2)$
$F(1, -2)$	$F(1, -1)$	$F(1, 0)$	$F(1, 1)$	$F(1, 2)$

Table 2.7 Values of the pixels following an image filter

$P(i-1, j-2)$	$P(i-1, j-1)$	$P(i-1, j)$	$P(i-1, j+1)$	$P(i-1, j+2)$
$P(i, j-2)$	$P(i, j-1)$	$P(i, j)$	$P(i, j+1)$	$P(i, j+2)$
$P(i+1, j-2)$	$P(i+1, j-1)$	$P(i+1, j)$	$P(i+1, j+1)$	$P(i+1, j+2)$

2.4.1 Concept of Frequency in Images

The image frequency can be defined as changing in gray level values with distance. The high-frequency coefficients are defined by large changes in the gray level value over a small distance, i.e., noise and edges. The low-frequency coefficients are defined by small changes in the gray level value over a large distance, i.e., background. Based on these frequencies, two types of spatial filter—the low-pass filter and high-pass filter—are used for image enhancement in the spatial domain. The high-pass filter passes high-frequency coefficients and removes low-frequency coefficients. This filter removes background and skin texture information from the image. The low-pass filter passes low-frequency coefficients and removes high-frequency coefficients. This filter removes noise and edges from the image. The low-pass filter is known as the smoothing filter and the high-pass filter is known as the sharping filter.

2.4.2 Low-Pass Average Filter

The idea of low-pass average filtering is that it replaces the gray level value of every pixel with the average gray level value of its neighborhood pixels. The resultant image has less sharp transitions in the gray level. Random noise has sharp transitions in gray levels, so this filter can be used for smoothing of an image. The two types of averaging mask—the standard average filter and weighted average filter—are used for image smoothing (as shown in Fig. 2.23).

The SCILAB code for the low-pass average filter is given here and the output of this code is shown in Fig. 2.24.

SCILAB Code for the Low-Pass Average Filter

```
a=uigetfile('*.*','Select the Image:-');
a=imread(a);
b=double(a);
c=imnoise(a,'salt & pepper',0.2);
d=double(c);

m=(1/9)*(ones(3,3));
[r1,c1]=size(a);

for i=1:r1
  for j=1:c1
    new(i,j)=a(i,j);
  end
end

for i=2:1:r1-1
  for j=2:1:c1-1

    new(i,j)=(m(1)*d(i-1,j-1))+(m(2)*d(i-1,j))+(m(3)*d(i-1,j+1))
    ......
          +(m(4)*d(i,j-1))+(m(5)*d(i,j))+(m(6)*d(i,j+1))......
          +(m(7)*d(i+1,j-1))+(m(8)*d(i+1,j))+(m(9)*d(i+1,j+1));

  end
end
imshow(uint8(new));
```

Fig. 2.23 Averaging filter masks: (**a**) standard; and (**b**) weighted

a

1/9	1/9	1/9
1/9	1/9	1/9
1/9	1/9	1/9

b

1/16	2/16	1/16
2/16	4/16	1/16
1/16	2/16	1/16

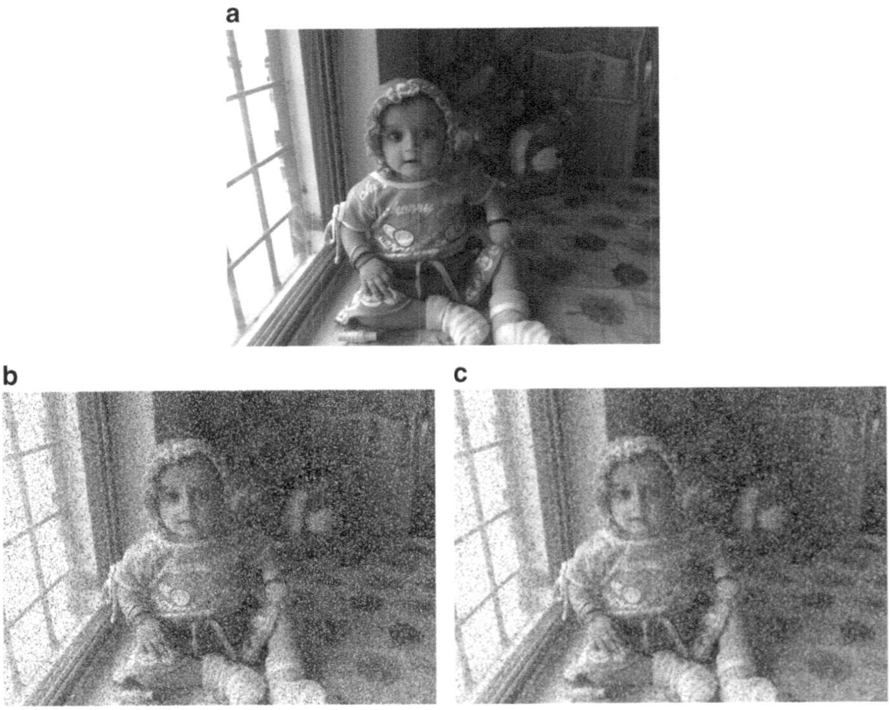

Fig. 2.24 Result of low-pass average filtering: (**a**) original image; (**b**) image corrupted by salt and pepper noise; and (**c**) filtered image

2.4.3 Low-Pass Median Filter (Order Statistic Filter)

We can see from Fig. 2.24 that the low-pass average filter removes noise by blurring it until it is no longer seen. But this process also blurs the edges. Hence, by using an averaging mask, more blurring can be available in the filtered image. So, rather than an averaging process, a non-linear filter known as a median filter is used. This is also known as an order statistics filter because its response is based on the ordering or ranking of the pixels contained within the mask. The steps for this filter are as follows:

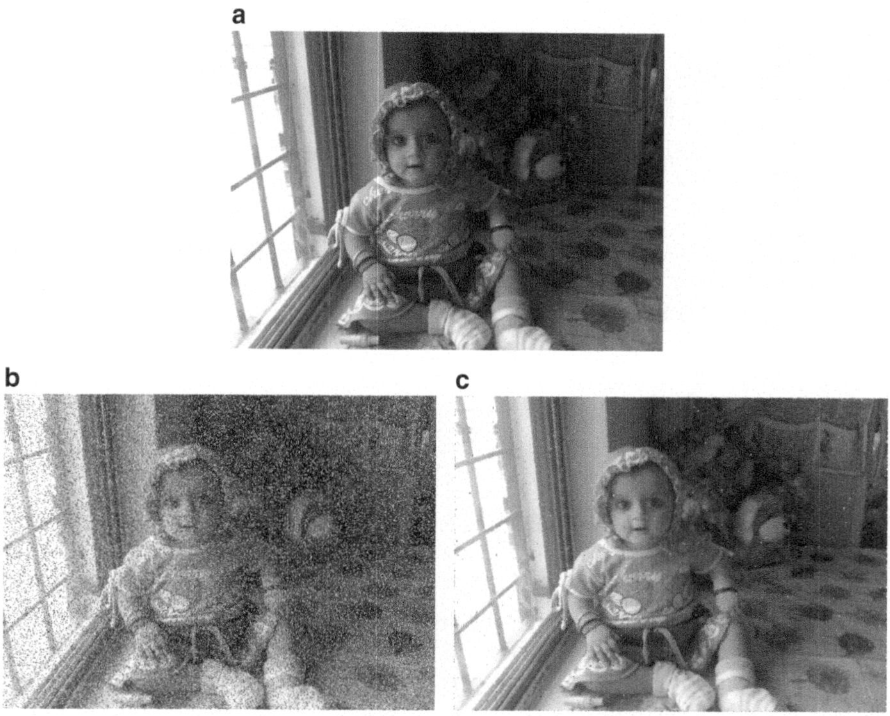

Fig. 2.25 Result of low-pass median filtering: (**a**) original image; (**b**) image corrupted by salt and pepper noise; and (**c**) filtered image

- Assume a 3 × 3 empty matrix.
- Place the mask at the left-hand corner.
- Arrange nine pixels in ascending or descending order.
- Choose the middle (median) value of the nine values.
- Place the value at the center.
- Move the mask in a similar fashion to the averaging filter until the end of the image.

The SCILAB code for the low-pass median filter is given here and the output of this code is shown in Fig. 2.25.

SCILAB Code for the Low-Pass Median Filter

```
a=uigetfile('*.*','Select the Image:-');
a=imread(a);
b=double(a);
c=imnoise(a,'salt & pepper',0.2);
d=double(c);


m=(1/9)*(ones(3,3));
[r1,c1]=size(a);
```

```
[r1,c1]=size(a);
for i=2:r1-1
  for j=2:c1-1
    a1=[d(i-1,j-1) d(i-1,j) d(i-1,j+1) d(i,j-1) d(i,j) d(i,j+1) d
(i+1,j-1) d(i+1,j) d(i+1,j+1)];
    a2=gsort(a1);
    med=a2(5);
    b(i,j)=med;
  end
end

imshow(uint8(b));
```

2.4.4 High-Pass Filter

The principle objective of sharpening is to highlight fine details or enhance details that have been blurred. The 3 × 3 mask used for high-pass filtering is shown in Table 2.8.

As we can see, the sum of the elements of the high-pass mask must equal zero. This is because when we place the mask on the low-frequency region (i.e., over an area of constant gray level), the result must be zero. The filter reduces the average gray level of the image to zero. This means that there are some negative gray levels in the output of the filter. Thus, some form of scaling is required.

The SCILAB code for the high-pass filter is given here and the output of this code is shown in Fig. 2.26.

Table 2.8 3 × 3 mask used for high-pass filtering

−1	−1	−1
−1	8	−1
−1	−1	−1

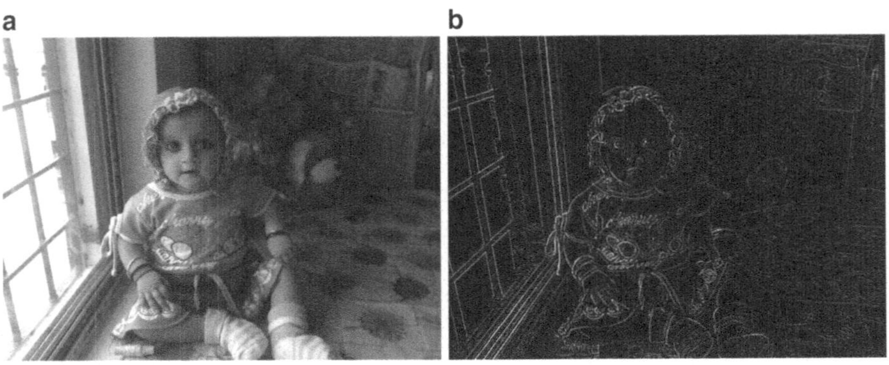

Fig. 2.26 Result of high-pass filtering: (**a**) original image; and (**b**) filtered image

SCILAB Code for High Pass Filter

```
a=uigetfile('*.*','Select the Image:-');
a=imread(a);
d=double(a);
[r1,c1]=size(a);
m=[-1 -1 -1;-1 8 -1;-1 -1 -1];
for i=2:1:r1-1
  for j=2:1:c1-1
   new(i,j)=(m(1)*d(i-1,j-1))+(m(2)*d(i-1,j))+(m(3)*d(i-1,j+1))
   ......
         +(m(4)*d(i,j-1))+(m(5)*d(i,j))+(m(6)*d(i,j+1)).......
         +(m(7)*d(i+1,j-1))+(m(8)*d(i+1,j))+(m(9)*d(i+1,j+1));
  end
end
imshow(uint8(new));
```

2.4.5 *High-Boost Filter*

We can observe from Fig. 2.25 that the high-pass filter gets rid of the complete background. So, to enhance the edges and also retain the background we use a modified version of the high-pass filter known as a high-boost filter. It is calculated as the difference between an original image and the low-pass filtered image.

$$\text{High Pass} = \text{Original} - \text{Low Pass} \tag{2.24}$$

Multiplying the original image by an amplification factor A gives the definition of a high-boost filter.

$$\begin{aligned}\text{High Boost} &= A \times (\text{Original}) - \text{Low Pass}\\ &= (A-1) \times (\text{Original}) + \text{Original} - \text{Low Pass}\\ &= (A-1) \times (\text{Original}) + \text{High Pass}\end{aligned} \tag{2.25}$$

For $A = 1$, it gives a simple high-pass filter. For $A > 1$, a fraction of the original image is added back to the filtered image, which gives an image similar to the original image. The process of subtracting a low-pass filtered image from the original image is called unsharp masking. The sample mask for this filter is shown in Table 2.9.

The SCILAB code for the high-boost filter is given here and the output of this code is shown in Fig. 2.27.

Table 2.9 Sample mask for high-pass filtering

−1/9	−1/9	−1/9
−1/9	$W = 9A - 1$	−1/9
−1/9	−1/9	−1/9

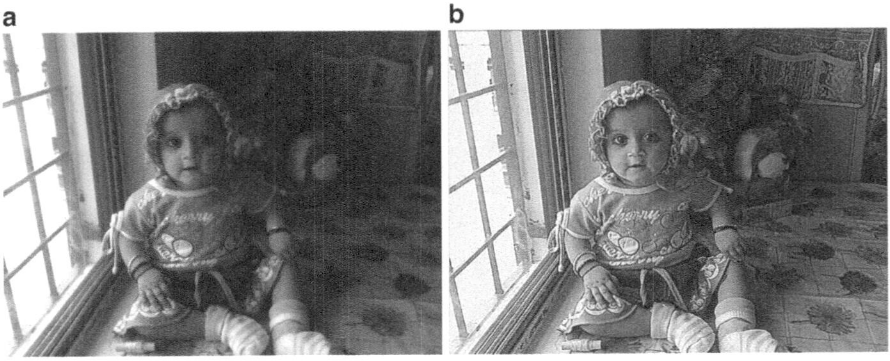

Fig. 2.27 Result of high-boost filtering: (**a**) original image; and (**b**) filtered image

SCILAB Code for High Boost Filter

```
a=uigetfile('*.*','Select the Image:-');
a=imread(a);
d=double(a);
[r1,c1]=size(a);
A=1.1;
BB=(9*A)-1;
m1=[-1 -1 -1;-1 BB -1;-1 -1 -1];
m=[-1 -1 -1;-1 8 -1;-1 -1 -1];
for i=2:1:r1-1
  for j=2:1:c1-1
    new(i,j)=(m1(1)*d(i-1,j-1))+(m1(2)*d(i-1,j))+(m1(3)*d(i-1,
j+1)).....
        +(m1(4)*d(i,j-1))+(m1(5)*d(i,j))+(m1(6)*d(i,j+1)).....
        +(m1(7)*d(i+1,j-1))+(m1(8)*d(i+1,j))+(m1(9)*d(i+1,j+1));
  end
end
imshow(uint8(new));
```

2.5 Image Enhancement Using Arithmetic/Logical Operations

Arithmetic/logical operations are involved in image enhancement on a pixel by pixel level between two or more images. Logical operations such as AND, OR, or NOT (where the NOT operation is applied on a single image) are used for image enhancement. Arithmetic operations such as addition, subtraction, multiplication and division may be used as and when required. The OR and AND operations are mainly used in conjunction with morphological operations [1, 2]. The SCILAB code for logical operations is given here and the output of this code is shown in Fig. 2.28.

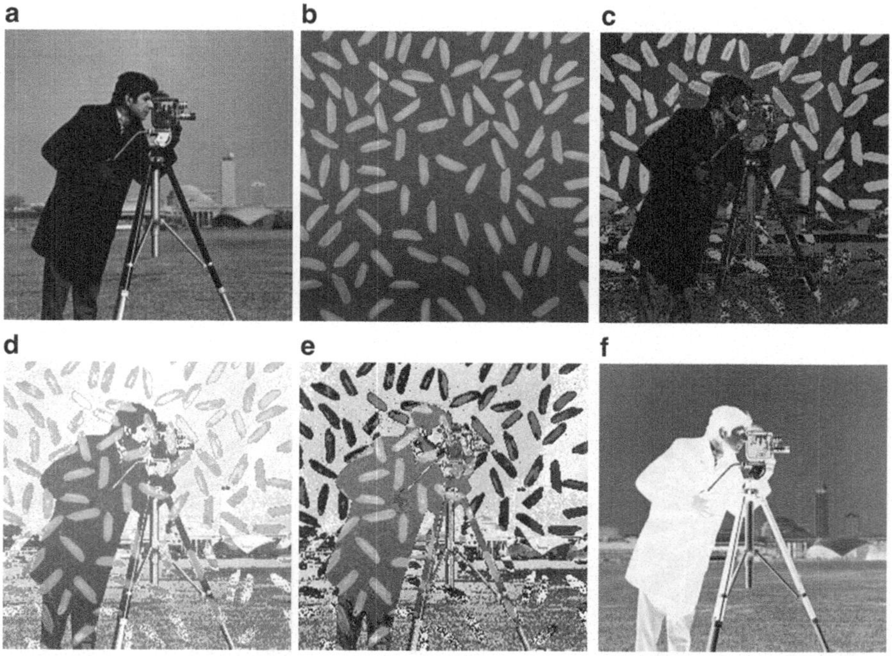

Fig. 2.28 Result of logical operations: (**a**) original image 1; (**b**) original image 2; (**c**) resultant image after AND operation; (**d**) resultant image after OR operation; (**e**) resultant image after Logical Exclusive ORing Operation (XOR) operation; and (**f**) resultant image after NOT operation

SCILAB Code for Logical Operations

```
a=uigetfile('*.*','Select the Image:-');
b=uigetfile('*.*','Select the Image:-');
C=bitand(A, B);
D=bitor(A, B);
E=bitxor(A, B);
F=bitcmp(A);
figure(1);
subplot(2,4,1);imshow(C);
subplot(2,4,2);imshow(D);
subplot(2,4,3);imshow(E);
subplot(2,4,4);imshow(F);
```

The SCILAB code for arithmetic operations is given here and the output of this code is shown in Fig. 2.29.

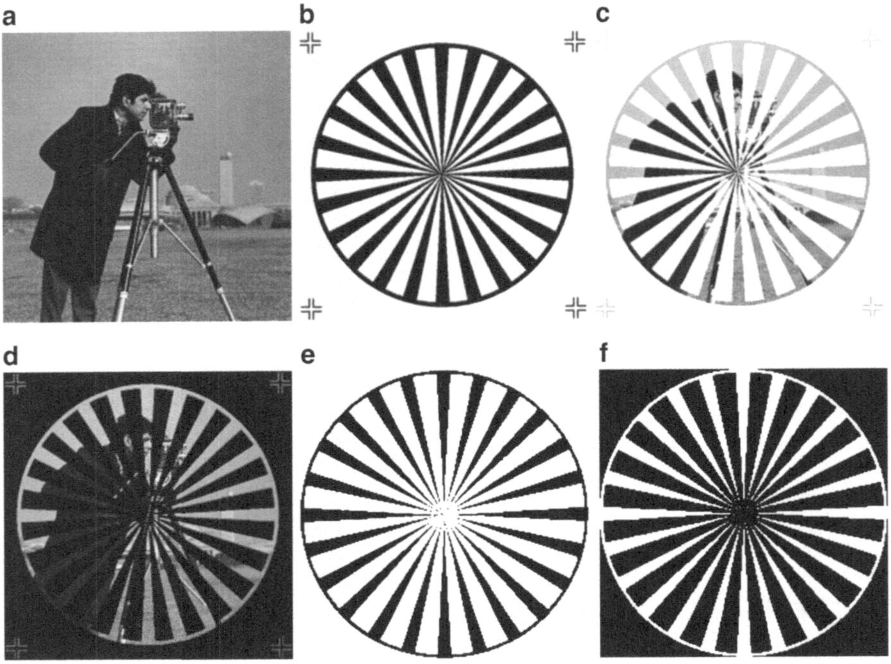

Fig. 2.29 Result of arithmetic operations: (**a**) original image 1; (**b**) original image 2; (**c**) resultant image after addition operation; (**d**) resultant image after subtract operation; (**e**) resultant image after multiplication operation; and (**f**) resultant image after division operation

SCILAB Code for Arithmetic Operations

```
a=uigetfile('*.*','Select the Image:-');
b=uigetfile('*.*','Select the Image:-');
C=imadd(A, B);
D=imsubtract(A, B);
E=immultiply(A, B);
F=imdivide(A);
figure(1);
subplot(2,4,1);imshow(C);
subplot(2,4,2);imshow(D);
subplot(2,4,3);imshow(E);
subplot(2,4,4);imshow(F);
```

Bibliography

1. Gonzalez, R., & Woods, R. (2008). *Digital image processing*. Upper Saddle River: Pearson Education India.
2. Gonzalez, R., Woods, R., & Eddins, L. (2009). *Digital image processing using MATLAB*. New Delhi: TATA McGraw-Hill Education.

Chapter 3
Image Enhancement in the Frequency Domain

3.1 Introduction

In Chap. 2, we studied what we mean by image enhancement and the way in which it is carried out in the spatial domain. In this chapter, we discuss image enhancement in the frequency domain. The theories of this type of image processing are beautifully described in the literature [1, 2].

Image enhancement in the frequency domain is processing the image in the Fourier domain. In spatial domain filtering, operation of the image is carried out by convolving the original image with a filter mask. Given the fundamentals of signal processing, filtering in the frequency domain would involve multiplying the image using the filter transfer function. The stepwise execution of filtering in the frequency domain can be given as follows:

- Multiply input image $f(x, y)$ by $(-1)^{(x + y)}$.
- Obtain $F(u, v)$ using the forward two-dimensional discrete Fourier transform (2D-DFT) of the result.
- Transform the image using the filter transfer function $H(u, v)$.

$$G(u, v) = U(u, v)F(u, v) \tag{3.1}$$

- Take the two-dimensional inverse discrete Fourier transform (2D-IDFT) of $G(u, v)$ and obtain part of the result.
- Multiply input image $f(x, y)$ by $(-1)^{(x + i)}$.

Before we discuss the procedures of frequency domain filtering, let us have a brief look at Fourier transform.

R. M. Thanki, A. M. Kothari, *Digital Image Processing using SCILAB*,
https://doi.org/10.1007/978-3-319-89533-8_3

3.2 Fourier Transform

The Fourier transform is of fundamental importance to image processing. It allows us to perform tasks which would be impossible to perform any other way; and its efficiency allows us to perform other tasks more quickly. The Fourier transform provides, among other things, a powerful alternative to linear spatial filtering; it is more efficient to use the Fourier transform than a spatial filter for a large filter. It also allows us to isolate and process particular image "frequencies", and so perform low-pass and high-pass filtering with a great degree of precision. The basic theory given by Fourier states that:

- Periodic functions can be represented by the sum of the sine/cosine functions of different frequencies, multiplied by a different coefficient.
- Non-periodic functions can be represented as the integral of sines/cosines multiplied by the weighing function.

Digital images are used as input data here. Thus, the discrete version of Fourier transform is used to convert a digital image into its frequency domain. This transform is known as discrete Fourier transform (DFT) and given as follows:

$$F(u,v)=\frac{1}{MN}\sum_{x=0}^{M-1}\sum_{y=0}^{N-1}f(x,y)e^{-j2\pi\left(\frac{ux}{M}+\frac{vy}{N}\right)} \tag{3.2}$$

where $u = 0, 1 \ldots M - 1$ and $v = 0, 1, 2 \ldots N - 1$.

Inverse DFT is given as follows:

$$f(x,y)=\sum_{u=0}^{M-1}\sum_{v=0}^{N-1}F(u,v)e^{j2\pi\left(\frac{ux}{M}+\frac{vy}{N}\right)} \tag{3.3}$$

where $x = 0, 1 \ldots M - 1$ and $y = 0, 1, 2 \ldots N - 1$.

The Fourier transform of a real function $f(x, y)$ is generally represented by complex values and defined as follows (see Fig. 3.1):

a b c

Fig. 3.1 (**a**) Original image and its Fourier transform; (**b**) magnitude response; and (**c**) phase response

$$F(u,v) = R(u,v) + jI(u,v) \tag{3.4}$$

where $F(u, v)$ is a Fourier transform of function, $R(u, v)$ is a real value, and $I(u, v)$ is a complex value.

The polar representation of function is given as follows:

$$F(u,v) = |F(u,v)|e^{j\varphi(u,v)} \tag{3.5}$$

where $\phi(u, v)$ is a phase response and $|F(u, v)|$ is a magnitude response.

The magnitude response and phase response of function is given as follows:

$$|F(u,v)| = \sqrt{R^2(u,v) + I^2(u,v)} \tag{3.6}$$

$$\varphi(u,v) = \tan^{-1}\frac{I(u,v)}{R(u,v)} \tag{3.7}$$

The SCILAB code for displaying fast Fourier transform (FFT) components of an image is given here and the result of the code is shown in Fig. 3.1.

SCILAB for Displaying Fast Fourier Transform Components

```
a=uigetfile('*.*','Select an Image');
a=imread(a);
a=double(a);
f=fft2(a);
m=abs(f);
p=angle(f);
subplot(131);
imshow(a);
title('Original Image');
subplot(132);
m=uint8(m);
imshow(m,[]);
title('Magnitude Plot');
subplot(133);
imshow(p,[]);
title('Phase Plot');
```

The frequency components produced by DFT can be arranged in an awkward manner, as shown in Fig. 3.2, which shows that that low-frequency components lie

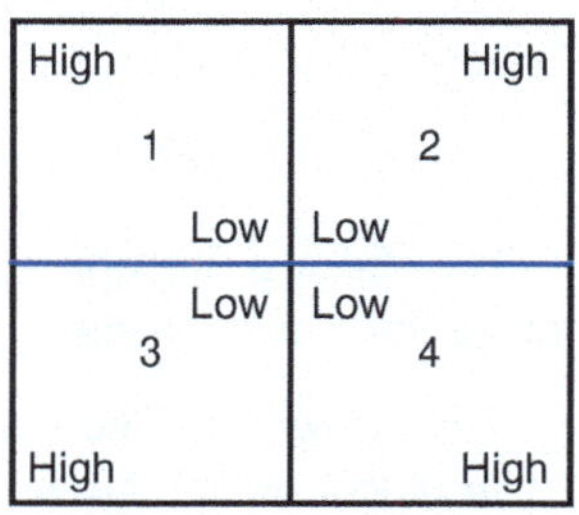

Fig. 3.2 Frequency distribution in discrete Fourier transform (DFT)

a b c

Fig. 3.3 Application of discrete Fourier transform (DFT): (**a**) original image; (**b**) low-frequency DFT components; and (**c**) high-frequency DFT components

in the center of the image and high-frequency components lie on the outside of the image. Figure 3.3 shows the low-frequency components and high-frequency components of the image after application of DFT.

3.2.1 *Important Properties of Discrete Fourier Transform (DFT)*

DFT has three very important properties: it is invariant against the shifting operation, scaling operation, and rotation operation. This transform is also called a rotation, scaling, and translation (RST) transform.

- *Shift Property*: The shift property is a basic operation that can convert an image in one domain to the other domain. Since both the spatial and frequency domain are periodic, shifting means rotating a component around the boundaries. The shifting property is calculated using the following equations:

$$F\left(f(x,y)e^{j2\pi\left(\frac{u_0 x}{M}+\frac{v_0 y}{N}\right)}\right) = F(u-u_0, v-v_0) \tag{3.8}$$

$$F(f(x-x_0, y-y_0)) = F(u,v)e^{-j2\pi\left(\frac{u x_0}{M}+\frac{v y_0}{N}\right)} \tag{3.9}$$

The shifting function shifts the phase of the image while the magnitude of the image remains unchanged.

- *Scale Property:* The scale property expands the frequency component values of an object, which grow in an image in the spatial domain. The scale property is calculated using the following equation:

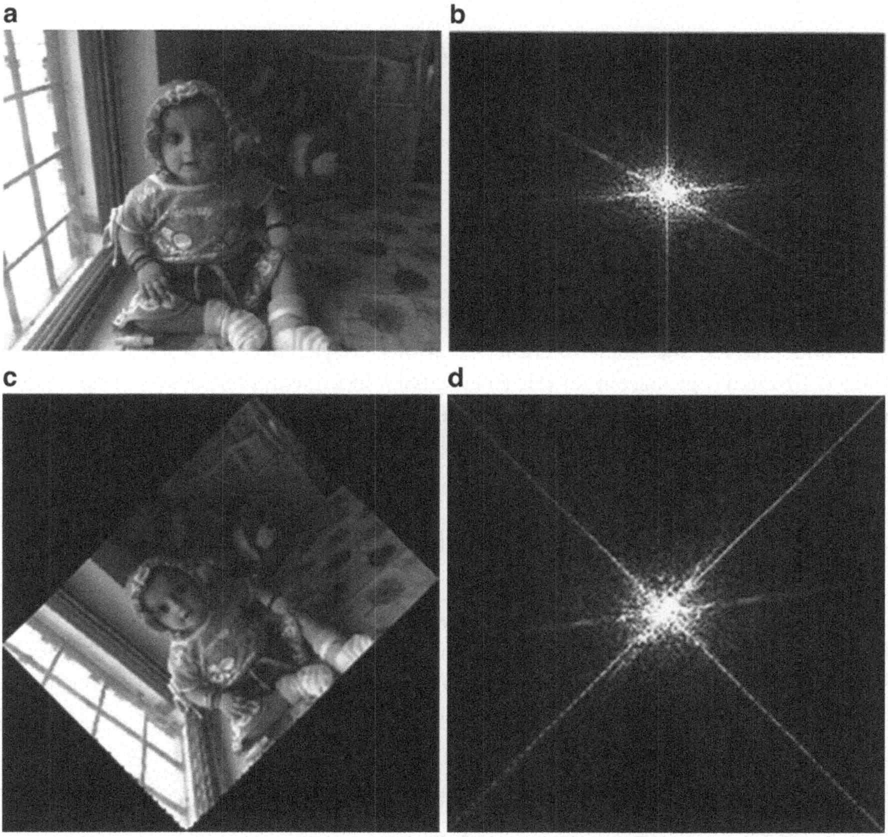

Fig. 3.4 Rotation property of discrete Fourier transform (DFT): (**a**) original image; (**b**) Fourier transform (FT) components of original image; (**c**) image rotated by 45° and (**d**) FT components of rotated image

$$F(f(ax, by)) = \frac{1}{|ab|}F\left(\frac{u}{a}, \frac{v}{b}\right) \tag{3.10}$$

- *Rotation Property*: The position of frequency components in the spatial location depends on the spatial domain content due to the separability of the DFT equation. The result is that if spatial domain values rotate, then the corresponding frequency component rotates. This property is depicted in Fig. 3.4. This property is defined using the following equation:

$$
\begin{aligned}
&F(f(x,y)) = F(u,v)\\
&F(f(x',y')) = F(u',v')\\
&\begin{bmatrix} x' \\ y' \end{bmatrix} = \begin{bmatrix} \cos\theta & -\sin\theta \\ \sin\theta & \cos\theta \end{bmatrix} \begin{bmatrix} x \\ y \end{bmatrix}\\
&\begin{bmatrix} u' \\ v' \end{bmatrix} = \begin{bmatrix} \cos\theta & -\sin\theta \\ \sin\theta & \cos\theta \end{bmatrix} \begin{bmatrix} u \\ v \end{bmatrix}
\end{aligned}
\tag{3.11}
$$

In this chapter, the filter function of various image filters such as the low-pass filter (LPF) and high-pass filter (HPF) are given along with their theory. Implementation of these filters in SCILAB is also described.

3.3 Low-Pass Frequency Domain Filter

The low-pass frequency domain filter allows low-frequency components to pass and high-frequency components are attenuated. The result is that any sharp transition such as edges and boundaries will be replaced with a smooth transition (blurred). This filter is known as the smoothing filter.

3.3.1 Ideal Low-Pass Filter (LPF)

The filter transfer function of an ideal LPF (ILPF) is given as follows:

$$
\begin{aligned}
H(u,v) &= 1, \text{for}\, D(u,v) \le D_0\\
H(u,v) &= 0, \text{otherwise}
\end{aligned}
\tag{3.12}
$$

where $D(u, v)$ is the distance of the frequency component (u, v) from the origin of the frequency plane and D_0 is the cut-off frequency which specified non-negative distance. Mathematically, $D(u, v)$ is given as follows:

$$
D(u,v) = \left[\left(u - \frac{M}{2}\right)^2 + \left(v - \frac{N}{2}\right)^2\right]^{\frac{1}{2}} \tag{3.13}
$$

This is called the ideal filter because it provides a sharp cut-off. All frequencies less than the cut-off are allowed to pass without attenuation, whereas all frequencies greater than the cut-off are completely attenuated. The strength of the filter can be calculated by computing the percentage of the total power contained within the circle of the radius of D_0 centered at the origin of the frequency plane. The power spectrum at (u, v) on the frequency plane is as follows:

$$P(u,v) = |F(u,v)|^2 = R^2(u,v) + I^2(u,v) \tag{3.14}$$

The total power is given as $P_T = \sum_{u=0}^{N-1} \sum_{v=0}^{N-1} P(u,v)$. So, the percentage of the total power contained with the circle of radius D_0 can be given as follows:

$$\beta = \left[\sum_u \sum_v P(u,v)/P_T \right] \times 100 \tag{3.15}$$

As the value of D_0 increases, the strength of the filter reduces, and vice versa. ILPF results in "rings" in the image. This means that ripples/waves are present whenever there are boundaries in an image.

The SCILAB code for the ILPF is given here and the result of the code is given in Fig. 3.5.

SCILAB for the Ideal LPF

```
a=uigetfile('*.*','Select the Image');
a=imread(a);
a=double(a);
r=size(a,1);
c=size(a,2);
d0=input('Enter the cut-off frequency - (Radius):-');
for u=1:1:r
  for v=1:1:c

    d=(((u-(r/2))^2)+((v-(c/2))^2))^0.5;

    if d<=d0
      h(u,v)=1;
    else
      h(u,v)=0;
    end

  end
end
b=fft2(a);
p_original=(abs(b))^2 + (atan(imag(b),real(b)))^2;
p_original=sum(sum(p_original))
c=fftshift(b);
c1=uint16(c);
new=c.*h;
new2=uint8(new);
new1=abs(ifft2(new));
imshow(uint8(new1));title(['Filtered Image with radius = ',
string(d0)]);
```

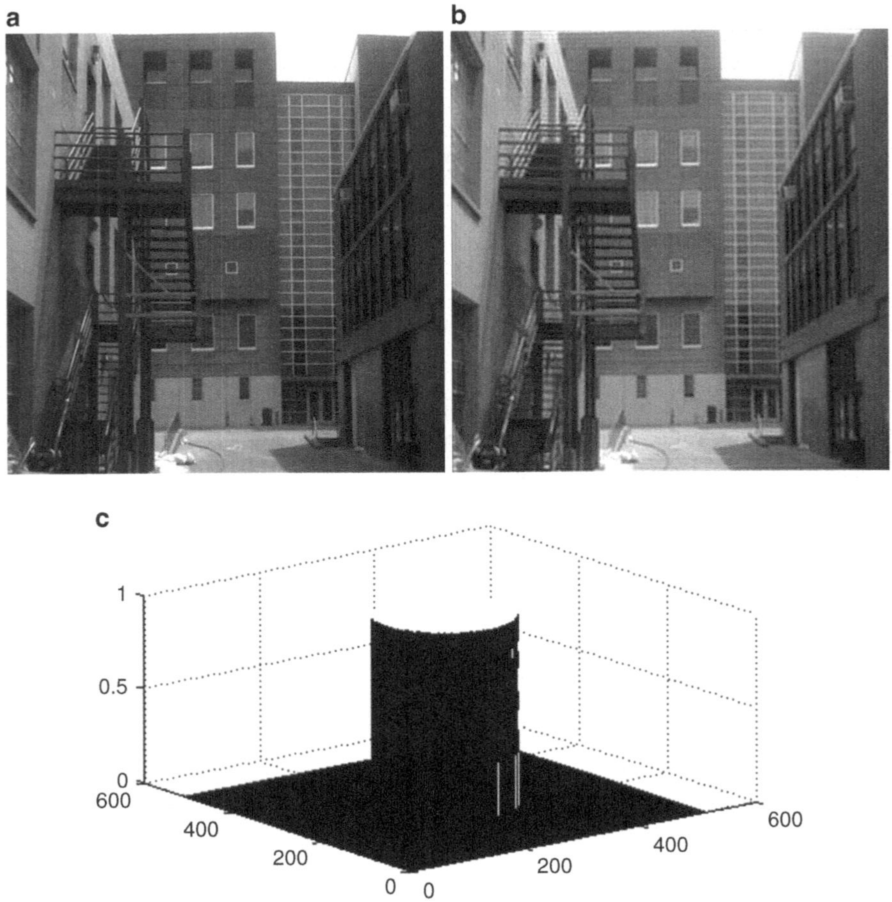

Fig. 3.5 Result of the ideal low-pass filter (LPF): (**a**) original image; (**b**) filtered image with cut-off frequency = 100; and (**c**) filter function response with radius = 100

3.3.2 Butterworth LPF

The transfer function of the Butterworth LPF of order n and with a cut-off frequency locus at a distance D_0 from the origin is defined as follows:

$$H(u,v) = \frac{1}{1 + \left[\frac{D(u,v)}{D_0}\right]^{2n}} \tag{3.16}$$

Unlike ILPF, the Butterworth LPF transfer function does not have a sharp discontinuity and clear cut-off between the passed frequency and filtered frequency. Normally, the filter function is defined such that when $D(u, v) = D_0$, $H(u, v)$ has maximum value. The commonly used $H(u, v)$ values for $D(u, v) = D_0$ are (1/2) and

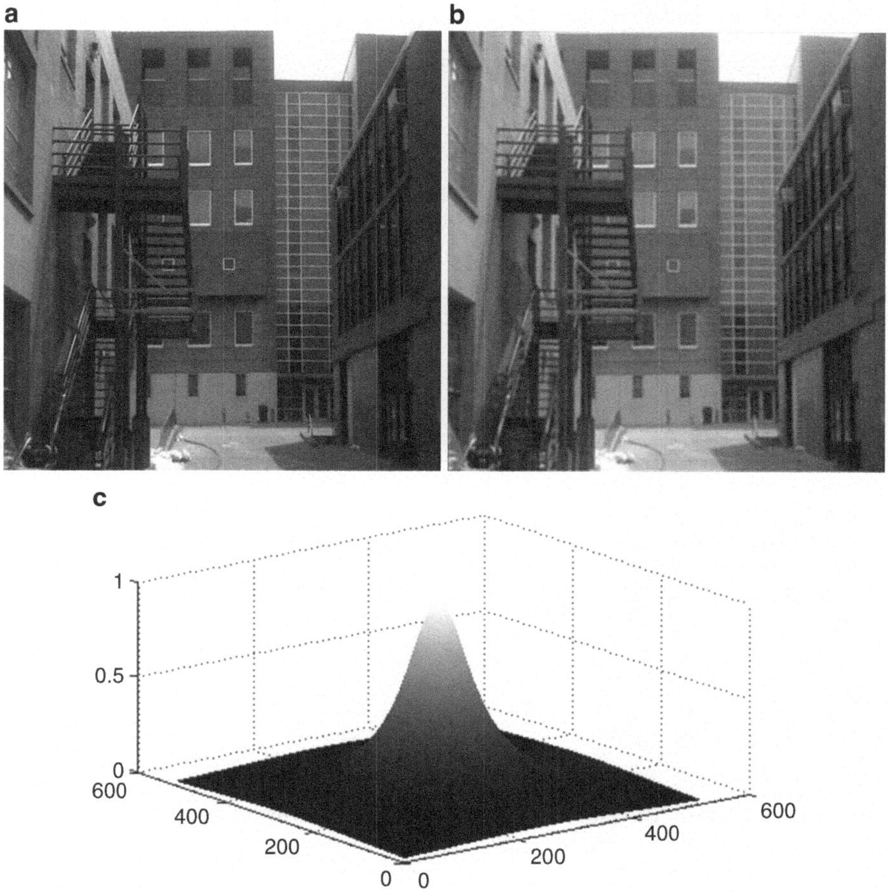

Fig. 3.6 Result of the Butterworth low-pass filter (LPF): (**a**) original image; (**b**) filtered image with cut-off frequency = 100 and order of filter = 3; and (**c**) filter function response with radius = 100 and order of filter = 3

$(1/\sqrt{2})$ of the maximum value. Equation 3.16 can be modified to yield the desired value of $H(u, v)$, which is $(1/\sqrt{2})$, down to its maximum value:

$$H(u,v) = \frac{1}{1 + 0.414[D(u,v)/D_0]^{2n}} \tag{3.17}$$

The SCILAB code for the Butterworth LPF is given here and the result of the code is given in Fig. 3.6.

SCILAB for the Butterworth LPF

```
a=uigetfile('*.*','Select the Image');
a=imread(a);
a=double(a);
r=size(a,1);
c=size(a,2);
d0=input('Enter the cut-off frequency - (Radius):-');
n=input('Enter the Order of the Filter:-');
for u=1:1:r
  for v=1:1:c
    d=(((u-(r/2))^2)+((v-(c/2))^2))^0.5;
    h(u,v)=1/(1+((d/d0)^(2*n)));
  end
end
b=fft2(a);
c=fftshift(b);
c1=c;
new=c.*h;
new2=new;
new1=ifft2(new2);
new1=abs(new1);
imshow(uint8(new1));title(['Filtered Image with radius and Order
of the Filter= ',string(d0),string(n)]);
```

3.3.3 *Gaussian LPF*

The transfer function of Gaussian LPF is defined as follows:

$$H(u,v) = e^{-D^2(u,v)/2\sigma^2} \tag{3.18}$$

The filter function is defined such that when $D(u, v) = D_0$, $H(u, v)$ has a maximum value of 0.607. The inverse FFT of Gaussian LPF is also Gaussian. The SCILAB code for Gaussian LPF is given here and the result of the code is given in Fig. 3.7.

SCILAB for Gaussian LPF

```
a=uigetfile('*.*','Select the Image');
a=imread(a);
a=double(a);
r=size(a,1);
c=size(a,2);
d0=input('Enter the cut-off frequency - (Radius):-');
for u=1:1:r
  for v=1:1:c
    d=(((u-(r/2))^2)+((v-(c/2))^2))^0.5;
    dd=d*d;
    h(u,v)=exp(-dd/(2*d0*d0));
  end
end
```

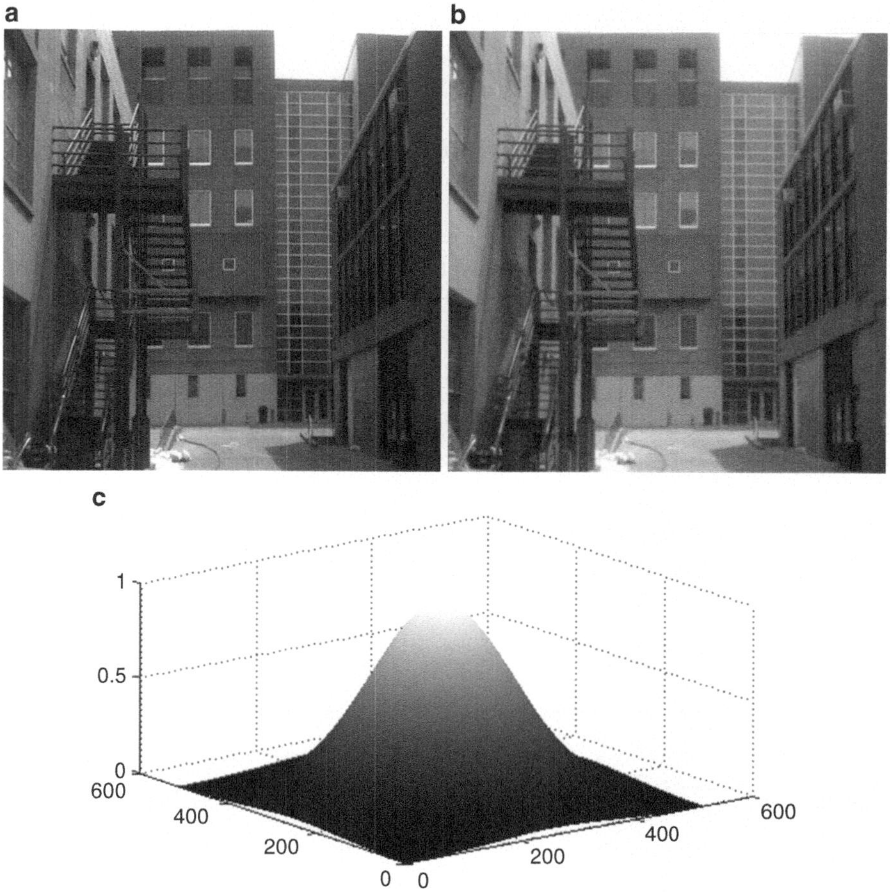

Fig. 3.7 Result of the Gaussian low-pass frequency (LPF): (**a**) original image; (**b**) filtered image with cut-off frequency = 100; and (**c**) filter function response with radius = 100

```
b=fft2(double(a));
c=fftshift(b);
c1=uint16(c);
new=c.*h;
new2=uint16(new);
new1=abs(ifft2(new));
imshow(uint8(new1));title(['Filtered Image with radius = ',
string(d0)]);
```

3.4 High-Pass Frequency Domain Filter

The high-pass frequency domain filter allows high-frequency components to pass and low-frequency components are attenuated. The result is that any smooth transition (blurred) in an image will be replaced with a sharp transition (edges). This filter is known as the sharpening filter.

3.4.1 Ideal High-Pass Filter (HPF)

The filter transfer function for the ideal HPF is given as follows:

$$\begin{aligned} H(u,v) &= 1, D(u,v) \geq D_0 \\ H(u,v) &= 0, \text{otherwise} \end{aligned} \tag{3.19}$$

It allows all high-frequency components with a distance from the origin of greater than D_0 without any filtering but completely attenuates all low frequency components. The SCILAB code for the ideal HPF is given here and the result of the code is given in Fig. 3.8.

SCILAB for Ideal HPF

```
a=uigetfile('*.*','Select the Image');
a=imread(a);
r=size(a,1);
c=size(a,2);
d0=input('Enter the cut-off frequency - (Radius):-');
for u=1:1:r
  for v=1:1:c

    d=(((u-(r/2))^2)+((v-(c/2))^2))^0.5;

    if d<=d0
      h(u,v)=0;
    else
      h(u,v)=1;
    end

  end
end
b=fft2(double(a));
c=fftshift(b);
c1=uint16(c);
new=c.*h;
new2=uint16(new);
new1=abs(ifft2(new));

imshow(uint8(new1));title(['Filtered Image with radius = ',
string(d0)]);
```

3.4.2 Butterworth HPF

The transfer function of the Butterworth HPF of order n and with a cut-off frequency locus at a distance D_0 from the origin is defined as follows:

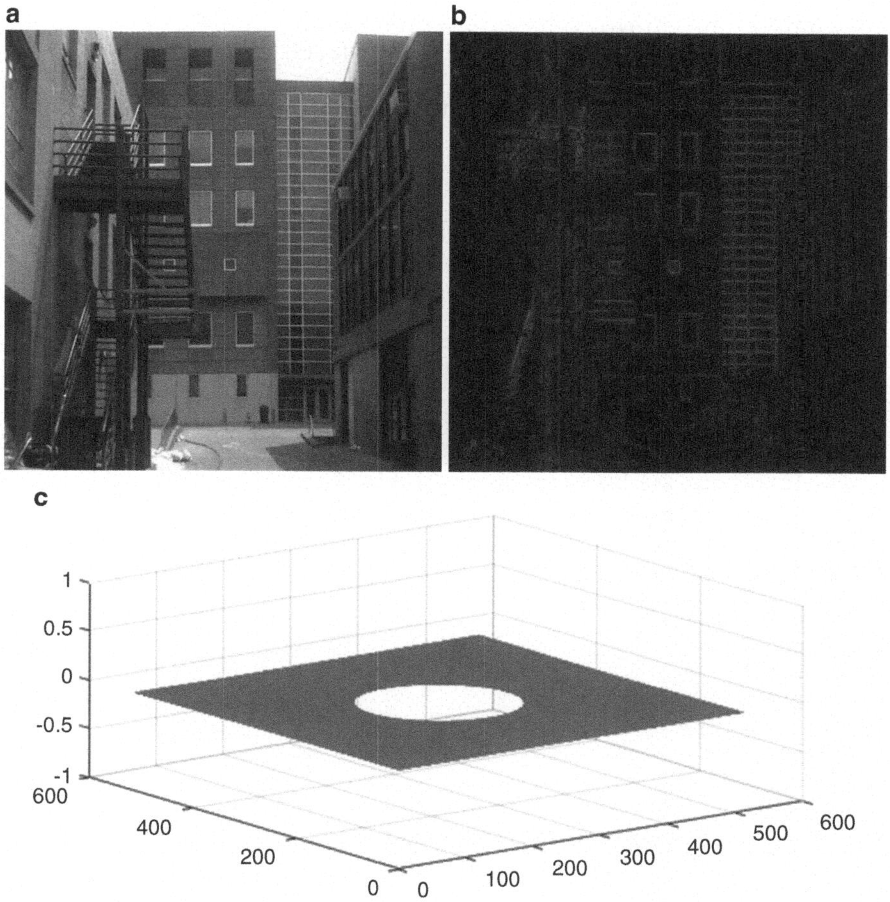

Fig. 3.8 Result of the ideal high-pass filter (HPF): (**a**) original image; (**b**) filtered image with cut-off frequency = 100; and (**c**) filter function response with radius = 100

$$H(u,v) = \frac{1}{1 + [D_0/D(u,v)]^{2n}} \tag{3.20}$$

Unlike ideal HPF, the Butterworth HPF transfer function does not have a sharp discontinuity and clear cut-off between the passed frequency and filtered frequency. Normally, the filter function is defined such that when $D(u, v) = D_0$, $H(u, v)$ is down to a certain fraction of its maximum value. The commonly used $H(u, v)$ values for $D(u, v) = D_0$ are (1/2) and $(1/\sqrt{2})$ of the maximum value. Equation 3.20 can be modified to yield the desired value of $H(u, v)$ which is $(1/\sqrt{2})$ of its maximum value:

$$H(u,v) = \frac{1}{1 + 0.414[D_0/D(u,v)]^{2n}} \tag{3.21}$$

The SCILAB code for the Butterworth HPF is given here and the result of the code is given in Fig. 3.9.

SCILAB for the Butterworth HPF

```
a=uigetfile('*.*','Select the Image');
a=imread(a);
r=size(a,1);
c=size(a,2);
d0=input('Enter the cut-off frequency - (Radius):-');
n=input('Enter the Order of the Filter:-');
for u=1:1:r
  for v=1:1:c

    d=(((u-(r/2))^2)+((v-(c/2))^2))^0.5;
    h(u,v)=1/(1+((d0/d)^(2*n)));

  end
end
b=fft2(double(a));
c=fftshift(b);
c1=uint16(c);
new=c.*h;
new2=uint16(new);
new1=abs(ifft2(new));

imshow(uint8(new1));title(['Filtered Image with radius and Order
of Filter = ',string(d0),string(n)]);
```

3.4.3 Gaussian HPF

The transfer function of the Gaussian HPF is defined as follows:

$$H(u,v) = 1 - e^{-D^2(u,v)/2D_0^2} \tag{3.22}$$

The SCILAB code for the Gaussian HPF is given here and the result of the code is given in Fig. 3.10.

SCILAB for the Gaussian HPF

```
a=uigetfile('*.*','Select the Image');
a=imread(a);
r=size(a,1);
c=size(a,2);
d0=input('Enter the cut-off frequency - (Radius):-');
for u=1:1:r
  for v=1:1:c
```

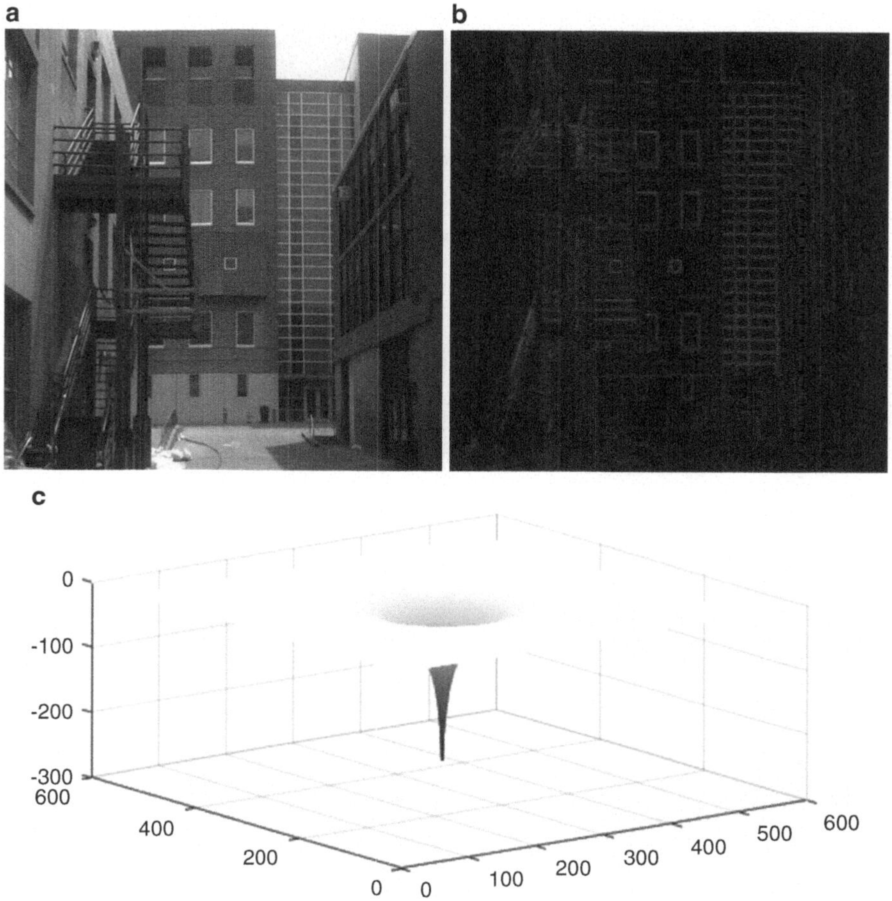

Fig. 3.9 Result of the Butterworth high-pass filter (HPF): (**a**) original image; (**b**) filtered image with cut-off frequency = 100; and (**c**) filter function response with radius = 100

```
        d=(((u-(r/2))^2)+((v-(c/2))^2))^0.5;
        dd=d*d;
        h(u,v)=1-exp(-dd/(2*d0*d0));

    end
  end
  b=fft2(double(a));
  c=fftshift(b);
  //c1=uint16(c);
  new=c.*h;
  new1=abs(ifft2(new));

  imshow(uint8(new1));title(['Filtered Image with radius and Crder
  of Filter = ',string(d0)]);
```

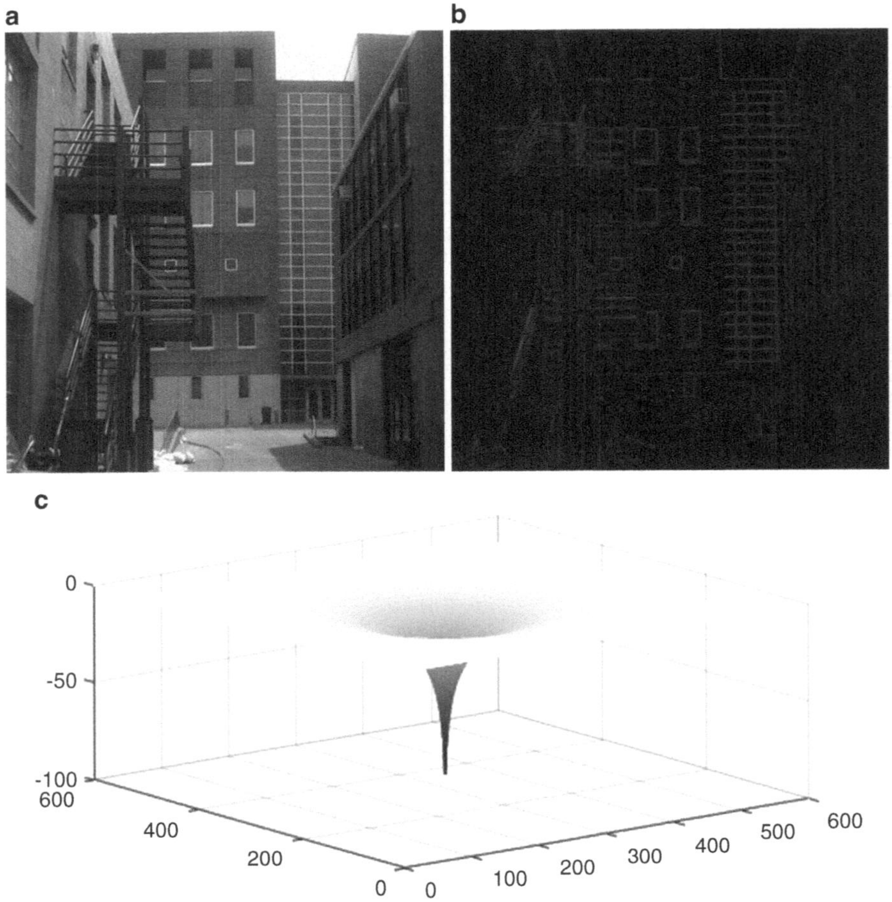

Fig. 3.10 Result of the Gaussian high-pass filter (HPF): (**a**) original image; (**b**) filtered image with cut-off frequency = 100; and (**c**) filter function response with radius = 100

3.5 Unsharp Masking

Like the spatial domain HPF, all regions with smooth gray levels are destroyed in the output image and only sharp transitions such as edges and boundaries remain dominant. To retain the regions with constant gray levels, a small positive constant is added to the filter transfer function before applying on $F(u, v)$. This makes certain sharp transitions in the output image more highlighted than they are in the original image. This concept is called high-frequency emphasis and is similar to the concept of unsharp masking. To obtain proper results, histogram equalization is applied to the output image.

3.5.1 Homomorphic Filtering

The gray level intensity $f(x, y)$ visible at any point (x, y) consists of two components: the amount of light incident on that point, called the illumination component $I(x, y)$, and the amount of light reflected from that point, called the reflectance component $r(x, y)$. So, $f(x, y)$ can be given as follows:

$$f(x, y) = i(x, y) \cdot r(x, y) \tag{3.23}$$

This equation is used as a basis because the illumination component contributes to smooth variations and thus to low-frequency components, whereas the reflectance component contributes to sharp transitions and thus to high frequency components in the frequency domain. Using Eq. 3.23, we can operate on low-frequency and high-frequency components simultaneously. But this equation cannot be used directly because 2D-DFT is not distributive over multiplication. That is, by taking a logarithm on both sides we can convert the multiplication into addition:

$$\begin{aligned} z(x, y) &= \ln[f(x, y)] \\ z(x, y) &= \ln[i(x, y)] + \ln[r(x, y)] \end{aligned} \tag{3.24}$$

Now 2D-DFT is given as follows:

$$Z(u, v) = I(u, v) + R(u, v) \tag{3.25}$$

Applying the transfer function:

$$S(u, v) = H(u, v) \cdot Z(u, v) \tag{3.26}$$

$$S(u, v) = H(u, v) \cdot I(u, v) + H(u, v) \cdot R(u, v) \tag{3.27}$$

Taking IDFT given as follows:

$$s(x, y) = i'(x, y) + r'(x, y) \tag{3.28}$$

Since $z(x, y) = \ln [f(x, y)]$, then the output image can be obtained as follows:

$$\begin{aligned} g(x, y) &= \exp[s(x, y)] \\ g(x, y) &= \exp[i'(x, y)]\exp[r'(x, y)] \\ g(x, y) &= i_0(x, y) \cdot r_0(x, y) \end{aligned} \tag{3.29}$$

where $i_0(x, y) = \exp[i'(x, y)]$, $r_0(x, y) = \exp[r'(x, y)]$.

The SCILAB code for the homomorphic filter is given here and the result of the code is given in Fig. 3.11.

Fig. 3.11 Result of homomorphic filtering: (**a**) original image; (**b**) corrupted image; and (**c**) restored image

SCILAB for the Homomorphic Filter

```
I = uigetfile('*.*','Select the Image');
GM = im2double(I);
t = 512;
p = 1/t:1/t:1;
for k = 1:t
  l(k,:)=p(:);
end
product = GM.*l;
log_GM = log(GM + 1);
log_product = log(product + 1);
fft_GM = (fftshift(fft2(log_GM)));
```

```
fft_product = (fftshift(fft2(log_product)));
diff_ill = (fft_product - fft_GM);
restored = (fft_product - diff_ill);
restored_im = ifft2(restored);
restored_image = (exp(restored_im)) - 1;
abs_restored_image = abs(restored_image);
imshow(abs_restored_image);title('Restored Image');
```

Bibliography

1. Gonzalez, R., & Woods, R. (2008). *Digital image processing*. Upper Saddle River: Pearson Education India.
2. Gonzalez, R., Woods, R., & Eddins, L. (2009). *Digital image processing using MATLAB*. New Delhi: TATA McGraw-Hill Education.

Chapter 4
Image Restoration

4.1 Introduction

Images captured using an acquisition process do not possess exactly the same information as original object, as image has some degradation in acquisition process. This degradation can occur due to the process used to acquire the image and many acquisition sensors or devices. For example, in remote sensing and astronomy, images are degraded due to various atmospheric conditions, different lighting conditions in space, and the camera position of the satellite. The image degradations can be classified into several categories, among which point degradations (i.e., due to noise) and spatial degradations (i.e., due to blurring) commonly occur in many applications.

Image restoration is a process by which an image can be restored to its original form from its degraded version. By this definition the restoration process appears to be the same as image enhancement, but there are differences between these two processes, as listed here:

- Image enhancement is a subjective process, while image restoration is an objective process.
- Image enhancement procedures are designed to manipulate an image in order to take advantage of the psychophysical aspects of human visual system (HVS). Image restoration is oriented towards modeling the degradation and applying the inverse process in order to reconstruct an original image.
- Image enhancement cannot be measured by any quantitative parameters. Image restoration can be measured using model-dependent quantitative parameters.
- An example of image enhancement is contrast stretching. An example of image restoration is removing a blur in the image.

R. M. Thanki, A. M. Kothari, *Digital Image Processing using SCILAB*,
https://doi.org/10.1007/978-3-319-89533-8_4

4.2 Image Degradation and Restoration Process

Figure 4.1 shows the image degradation and restoration model. We can consider a noisy image to be modeled as follows:

$$g(x,y) = H[f(x,y)] + n(x,y) \tag{4.1}$$

where H is a degradation function, $f(x, y)$ is an original image, $n(x, y)$ is a noise parameter, and $g(x, y)$ is a noisy image.

The main objective of the image restoration process is to derive features of the original image from its degraded version $g(x, y)$ while having some knowledge of degradation function H and noise parameter $n(x, y)$. If H is a linear function with an invariant property, the degraded image in the spatial domain can be given as follows:

$$g(x,y) = h(x,y) \otimes f(x,y) + n(x,y) \tag{4.2}$$

where $h(x, y)$ represents the degradation function in the spatial domain, and $\otimes$ represents the convolution.

Using the knowledge of the relationship between the spatial and frequency domain, Eq. 4.2 can be write in the frequency domain as follows:

$$G(u,v) = H(u,v) \cdot F(u,v) + N(u,v) \tag{4.3}$$

The degradation H function is linear and it is invariant can be modeled as convolution. Therefore, the restoration process is also sometimes referred as the deconvolution process. In Sect. 4.3, the function H is an identity operator and image degradation is due to additive noise.

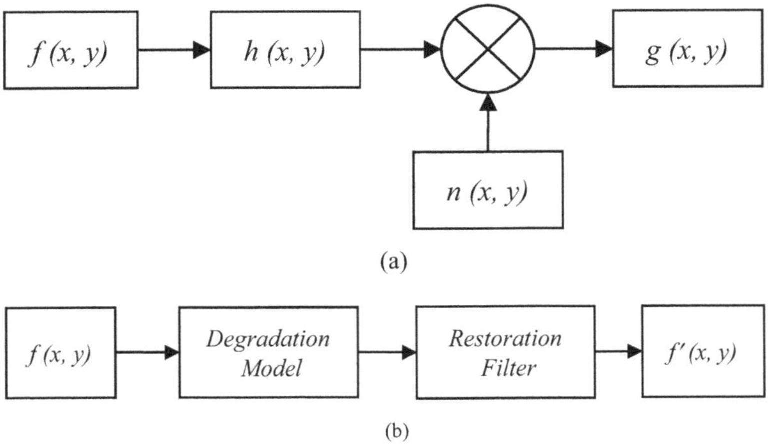

Fig. 4.1 (**a**) Image degradation model. (**b**) Image restoration model

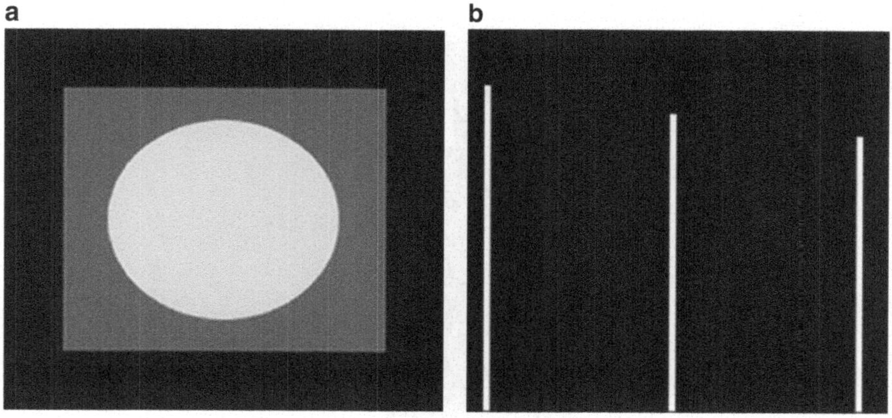

Fig. 4.2 Test pattern images

4.3 Noise Models

Most images are degraded by noise during their acquisition and/or transmission. Many things may affect the hardware performance of image acquisition/transmission tools. For example, the lighting and temperature of the image sensors in the charge-coupled device (CCD) camera are two major parameters which generate noise in an acquired image. Researchers assume that noise is an independent function of spatial coordinates and it is uncorrelated to the image. The spatial noise descriptor gives the statistical behavior of noise intensity. This intensity can be considered as a random variable and is given by the probability density function (PDF). There are six main types of noise sources that degrade images, which are explained in the following sections. The concept of generation of noise PDFs is beautifully described by Gonzales and Wood [1, 2]. We have used the test pattern shown in Fig. 4.2 to show the effect of various noises.

4.3.1 Gaussian Noise

The Gaussian noise PDF is given as follows:

$$p(n) = \frac{1}{\sqrt{2\pi}\sigma} e^{-(n-\mu)^2/2\sigma^2} \tag{4.4}$$

where n is the noise intensity, μ is mean value of n, σ is standard deviation, and σ^2 is the variance of n.

Examples of Gaussian noise are noise due to electronic circuits and sensors. The histogram of the Gaussian noise PDF and its effect on images is given as Fig. 4.3. In

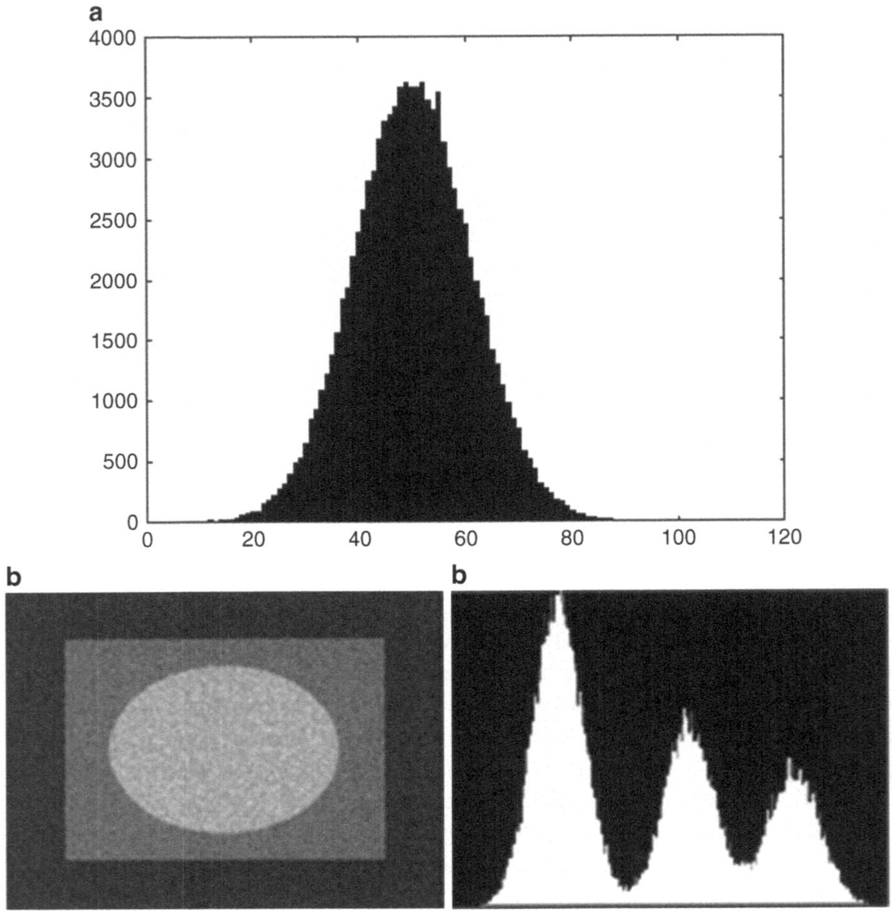

Fig. 4.3 (**a**) Gaussian noise probability density function (PDF). (**b**, **c**) Noisy images with Gaussian noise

SCILAB, inbuilt functions such as *imnoise* are available which add different types of noise in the image. The SCILAB code for generation of Gaussian noise PDF is as follows:

```
a=0;b=1; c=10000;d=1;
r=a+b*rand(c,d);
bins=100;
p=imhist(r,bins);
plot(p);
```

The SCILAB code for adding Gaussian noise in an image is as follows and the results are shown in Fig. 4.3:

```
a=uigetfile('*.*','Select the Image:-');
a=imread(a);
b=imnoise(a,'gaussian');
imshow(b);
```

4.3.2 Rayleigh Noise

The Rayleigh noise PDF is given as follows:

$$
\begin{aligned}
p(n) &= \left(\frac{2}{b}\right)(n-a)e^{-(n-a)^{\frac{2}{b}}}, n \geq a \\
p(n) &= 0, \text{otherwise}
\end{aligned}
\tag{4.5}
$$

Here, $\mu = a + \sqrt{\pi b/4}$ and $\sigma^2 = \frac{b(4-\mu\ )}{4.}$.

This equation is used for approximation of the histogram of images. The histogram of Rayleigh noise PDF and its effect on images are given in Fig. 4.4. The SCILAB code for generation of Rayleigh noise PDF as follows:

```
a=0;b=1; c=10000;d=1;
r1=a+(-b*log(1-rand(c,d))).^0.5;
bins=100;
p1=imhist(r1,bins);
plot(p1);
```

4.3.3 Erlang/Gamma Noise

The Erlang noise PDF is given as follows:

$$
\begin{aligned}
p(n) &= \frac{a^b n^{b-1}}{(b-1)!}, n \geq 0 \\
p(n) &= 0, \text{otherwise}
\end{aligned}
\tag{4.6}
$$

Here, $\mu = \frac{b}{a}, \sigma^2 \ = \frac{b}{a^2}$ and $a > 0$, and b is a positive integer.

The histogram of the Erlang noise PDF and its effect on images is given in Fig. 4.5. The SCILAB code for generation of the Erlang noise PDF is as follows:

```
a=2;b=5; c=10000;d=1;
k1=-1/a;
R=zeros(c,d);
for j=1:b
  r2=R+k1*log(1-rand(c,d));
```

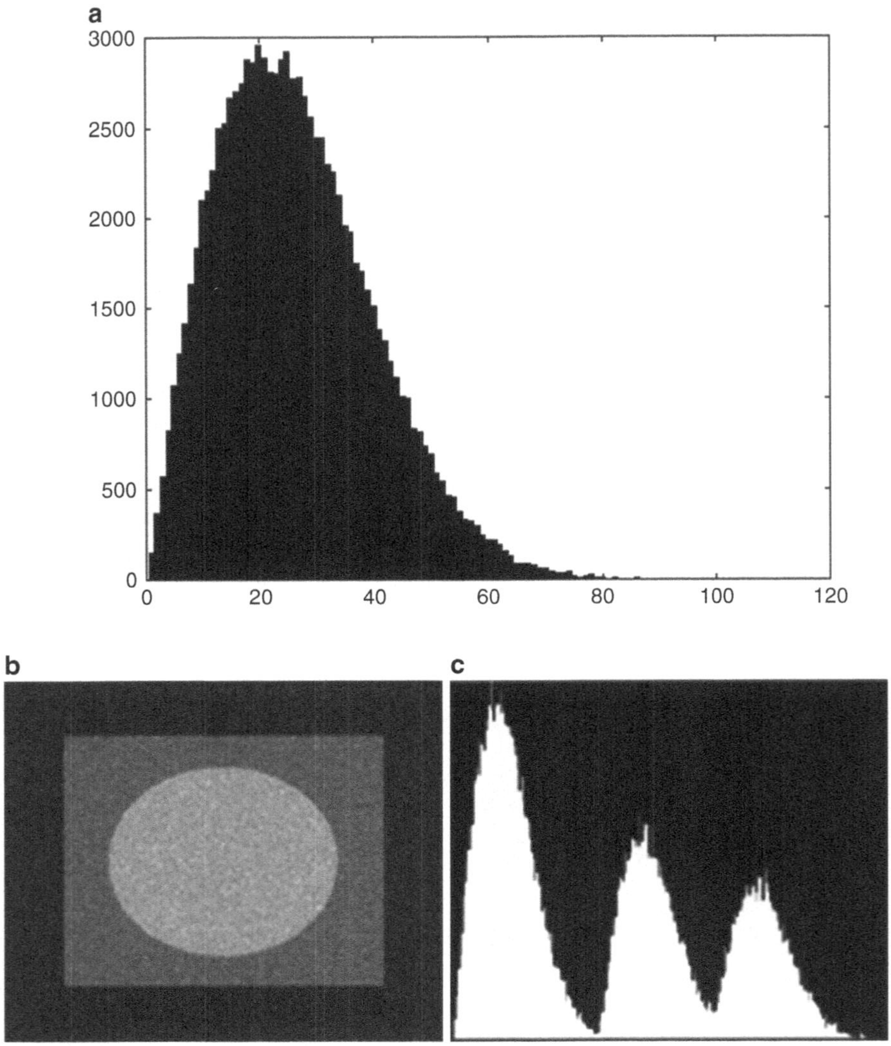

Fig. 4.4 (**a**) Rayleigh noise probability density function (PDF). (**b**, **c**) Noisy images with Rayleigh noise

```
end
bins=100;
p2=imhist(r2,bins);
plot(p2);
```

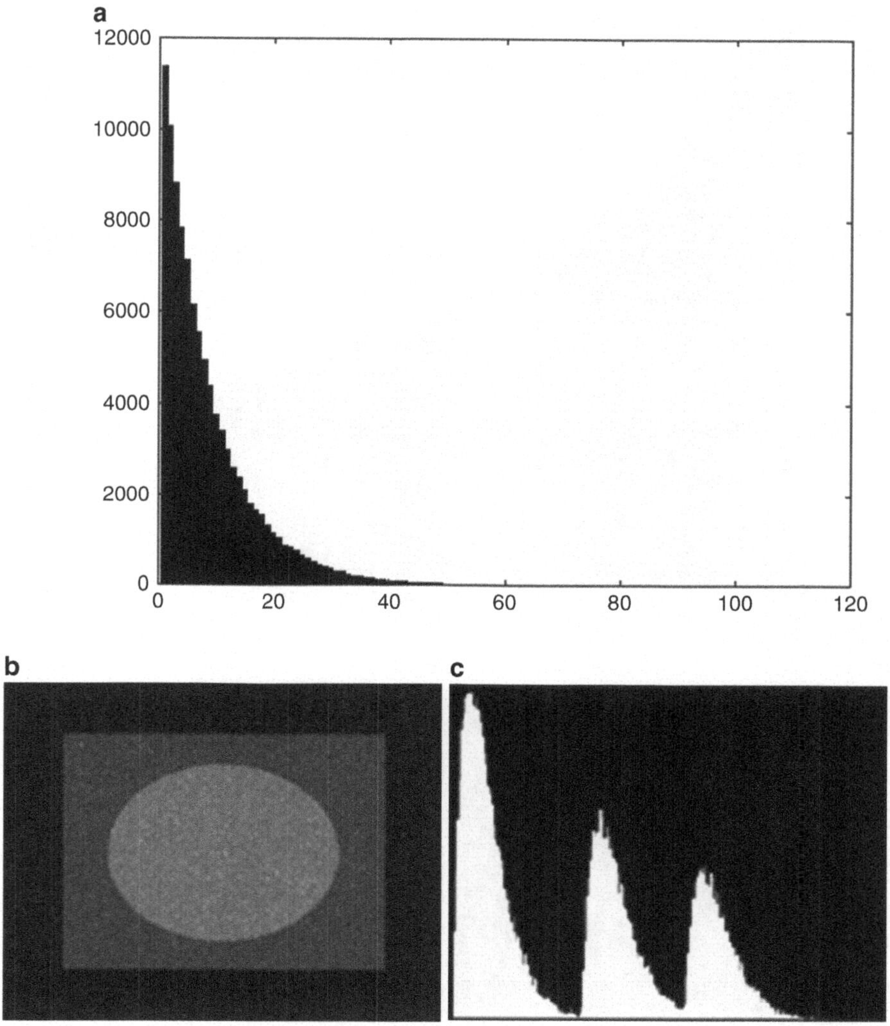

Fig. 4.5 (**a**) Erlang noise probability density function (PDF). (**b**, **c**) Noisy images with Erlang noise

4.3.4 Exponential Noise

Exponential noise is a special case of Erlang/gamma noise with $b = 1$. The noise PDF is given as follows:

$$\begin{aligned} p(n) &= a \cdot e^{-an}, n \geq 0 \\ p(n) &= 0, \text{otherwise} \end{aligned} \tag{4.7}$$

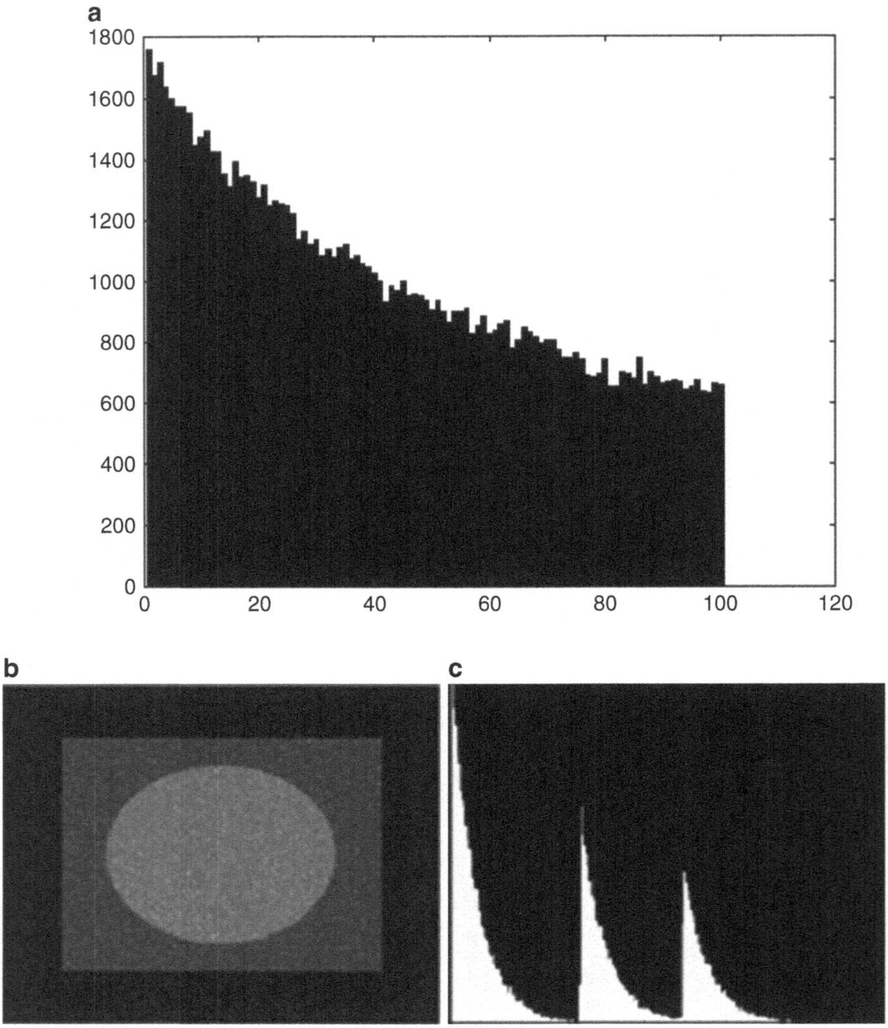

Fig. 4.6 (**a**) Exponential noise probability density function (PDF). (**b**, **c**) Noisy images by exponential noise

Here, $\mu = \frac{1}{a}, \sigma^2 = \frac{1}{a^2}$, $a > 0$. This type of noise is used in laser imaging. The histogram of the exponential noise PDF and its effect on an image are given in Fig. 4.6. The SCILAB code for generation of the exponential noise PDF is as follows:

```
a2=1; c=10000;d=1;
k2=1/a2;
r3=k2*exp(-a2*rand(c,d));
bins=100;
p3=imhist(r3,bins);
plot(p3);
```

4.3.5 *Uniform Noise*

The uniform noise PDF is given as follows:

$$\begin{aligned} p(n) &= \frac{1}{b-a}, a \le n \le b \\ p(n) &= 0, \text{otherwise} \end{aligned} \tag{4.8}$$

Here $\mu = \frac{a+b}{2}, \sigma^2 = \frac{(b-a)^2}{12}$.

The histogram of the uniform noise PDF and its effect on an image are given in Fig. 4.7. The SCILAB code for generation of uniform noise PDF is as follows:

```
a=0;b=1; c=10000;d=1;
r4=a+(b-a)*rand(c,d);
bins=100;
p4=imhist(r4,bins);
plot(p4);
```

4.3.6 *Salt and Pepper Noise*

The PDF of salt and pepper noise is given as follows:

$$\begin{aligned} p(n) &= Pa, n = a \\ p(n) &= pb, n = b \\ P(n) &= 0, \text{otherwise} \end{aligned} \tag{4.9}$$

If $b > a$, then the intensity of b will appear as a white dot on the image; it appears as a black dot if P is zero. The noise is called a dot. Frequently, Pa and Pb are saturated values, resulting in positive impulses being a white dot and negative impulses being a black dot. This noise shows up when quick transitions take place in the image. The histogram of the salt and pepper noise PDF and its effect on an image are shown in Fig. 4.8. The SCILAB code for generation of the salt and pepper noise PDF is as follows:

```
a=0;b=1; c1=1;d1=1;
X=rand(c1,d1);
n = find(X <= a);
r5(n) = 0;
u=a+b;
n = find(X > a & X <=u);
r5(n) = 1;
p5=imhist(r5);
plot(p5);
```

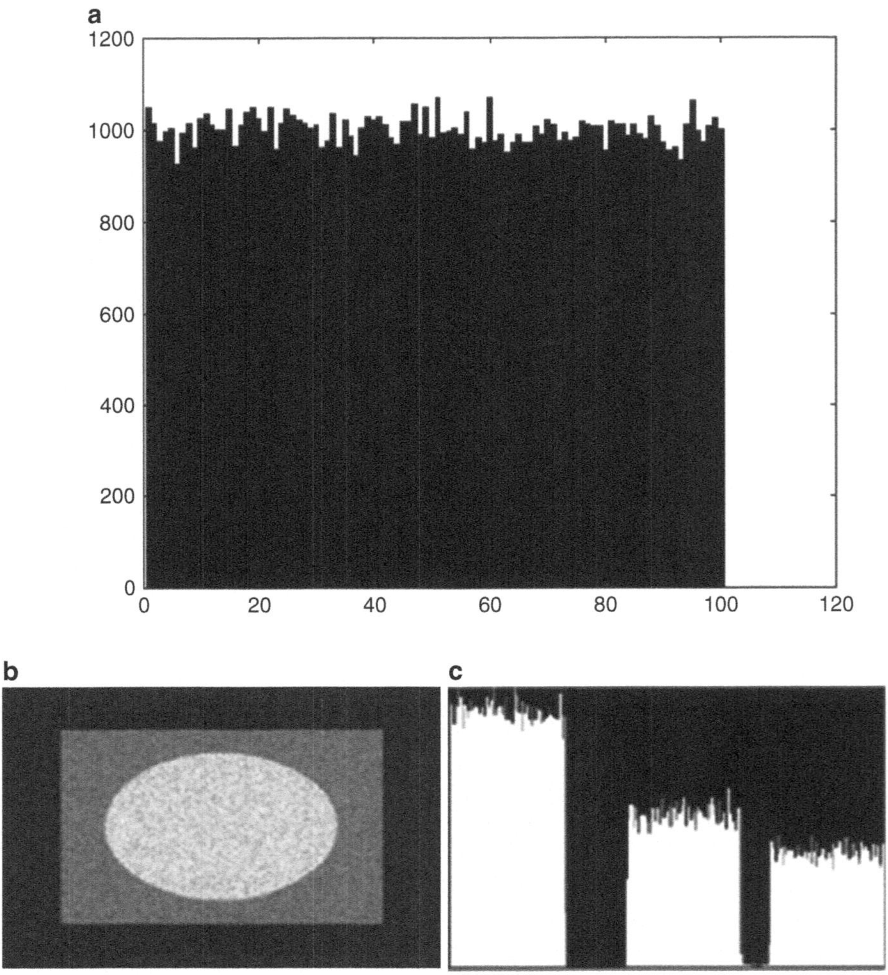

Fig. 4.7 (**a**) Uniform noise probability density function (PDF). (**b**, **c**) Noisy images with uniform noise

4.4 Periodic Noise and Estimation of Noise Parameters

Periodic noise appears in an image during its acquisition using electrical or electromechanical tools. This noise is a space-dependent noise. If the strength of the noise is high it can be observed in an image by analyzing the image in the frequency domain.

The noise parameters of periodic noise are estimated in the frequency domain. The parameters of the noise PDF is partially derived from the specification of the acquisition tool. The noise parameters differ for different imaging systems. The easiest way to estimate noise parameters is from the set of acquired images from flat environments. The resulting images are usually the best indication of noise

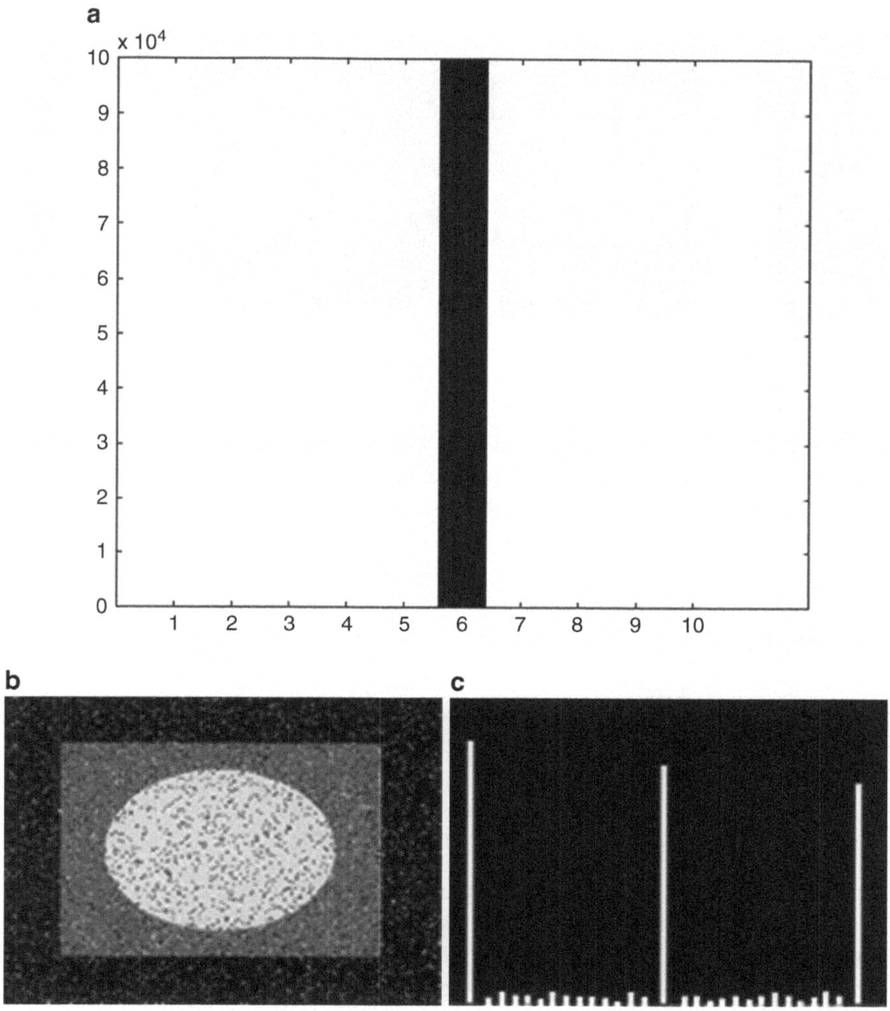

Fig. 4.8 (**a**) Salt and pepper noise probability density function (PDF). (**b**, **c**) Noisy images with salt and pepper noise

introduced by a system. Figure 4.9 shows back board images acquired and the noise presented in them.

The simplest use of the pixel values from the image strips is to calculate the mean and variance of noise intensity levels:

$$\mu = \sum_{ni \in S} ni \cdot p(ni)$$
$$\sigma^2 = \sum_{ni \in S} (ni - \mu)^2 p(ni) \tag{4.10}$$

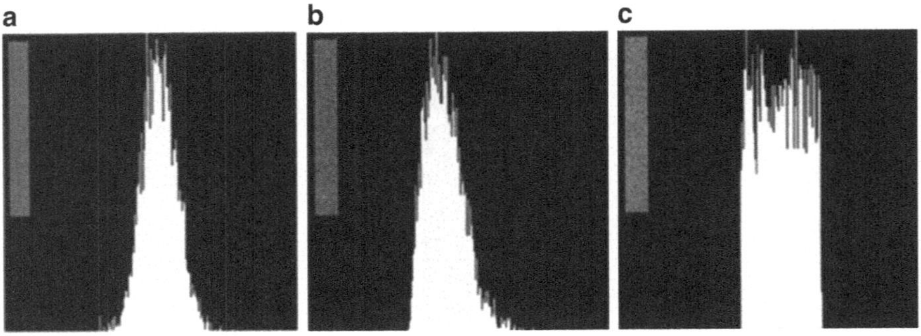

Fig. 4.9 Noisy back board images by (**a**) Gaussian noise; (**b**) Rayleigh noise; and (**c**) uniform noise

If the noise PDF shape is Gaussian in nature then the parameters of this noise are easily derived. For estimation of impulse noise, the probabilities of two intensity values, *Pa* and *Pb*, are derived from the histogram of the image.

4.5 Image Restoration: Spatial Filtering

If the image is degraded by only noise the degradation model for the image in the spatial domain and frequency domain is given as follows:

$$\begin{aligned} g(x,y) &= f(x,y) + n(x,y) \\ G(u,v) &= F(u,v) + N(u,v) \end{aligned} \tag{4.11}$$

In this case, different spatial filters are used to remove different kinds of noise. In Eq. 4.11, noise is an undefined term and its subtraction from the degraded image is a not a realistic option. Therefore, spatial filtering is a good candidate for removal of additive noise. We have already discussed spatial filters for image enhancement. Here, we discuss spatial filters for image restoration. Basically, spatial filters can be classified into two categories: noisy spatial filters (i.e., mean filters and order statistics filters) and adaptive spatial filters.

4.5.1 Mean Filters

The various types of mean filters applied on these pixels are given in Sects. 4.5.1.1, 4.5.1.2, 4.5.1.3, and 4.5.1.4. Here, P_{xy} represents the set of pixel coordinates with size of $M \times N$, centered at the point (x, y).

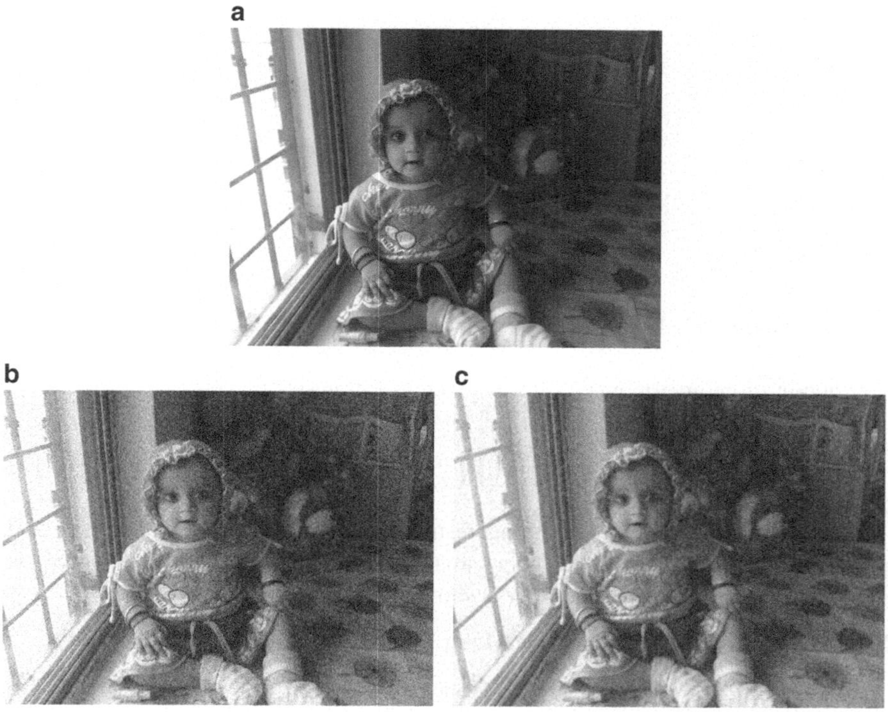

Fig. 4.10 Result from the arithmetic mean filter: (a) original image; (b) noisy image; and (c) filtered image

4.5.1.1 Arithmetic Mean Filter

The arithmetic mean filter calculates the average value of a degraded image $g(x, y)$ in the area defined by P_{xy}. The output of this filter is given as follows:

$$f'(x,y) = \frac{1}{MN}\sum\nolimits_{(p,t)\in P_{xy}} g(p,t) \qquad (4.12)$$

This filter removes blurring in the degraded image. The SCILAB code for the arithmetic mean filter is given here and the result of this filter is shown in Fig. 4.10.

SCILAB Code for the Arithmetic Mean Filter

```
a=uigetfile('*.*','Select the Image:-');
a=imread(a);
b1=double(a);
c=imnoise(a,'gaussian');
d=double(c);
b=d;
m=(1/9)*(ones(3,3));
[r1,c1]=size(a);
```

```
for i=2:r1-1
  for j=2:c1-1
    a1=d(i-1,j-1)+d(i-1,j)+d(i-1,j+1)+d(i,j-1)+d(i,j)+d(i,j+1)
+d(i+1,j-1)+d(i+1,j)+d(i+1,j+1);
    b(i,j)=a1*(1/9);
  end
end
imshow(uint8(b)); title('Filtered Image');
```

4.5.1.2 Geometric Mean Filter

The output of geometric mean filter is given as follows:

$$f'(x,y) = \left(\prod\nolimits_{(p,t)\in P_{xy}} g(p,t)\right)^{\frac{1}{MN}} \tag{4.13}$$

This filter gives better smoothing results than the arithmetic mean filter and provides more detail from the noisy image. The SCILAB code of this filter is given here and result from this filter is shown in Fig. 4.11.

SCILAB Code for the Geometric Mean Filter

```
a=uigetfile('*.*','Select the Image:-');
img=imread(a);
noise=imnoise(img,'gaussian',0,(100/(255)^2));
m=3;
n=3;
noise1=double(noise);
c=log(noise1);
F = imfilter(c,ones(m,n),'replicate');
f=exp(F).^(1/m/m);
K=f;
imshow(K);title('Filtered Image');
```

4.5.1.3 Harmonic Mean Filter

The output of the harmonic mean filter is given as follows:

$$f'(x,y) = \frac{MN}{\sum_{(p,t)\in P_{xy}} g(p,t)} \tag{4.13}$$

This filter works on all types of noisy images but does not work on images corrupted by pepper noise. The SCILAB code for this filter is given here and the result from this filter is shown in Fig. 4.12.

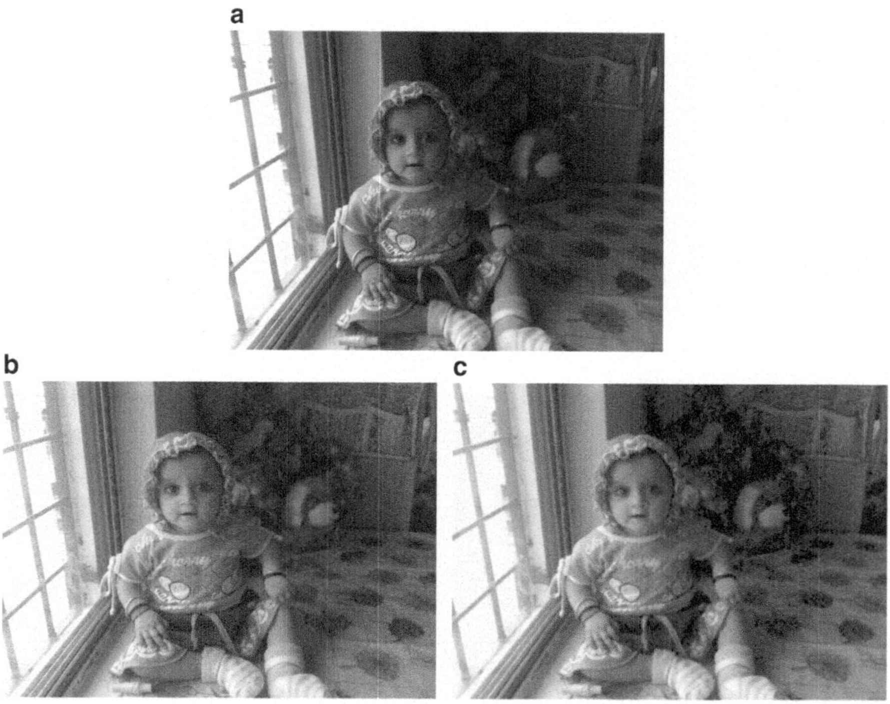

Fig. 4.11 Result from the geometric mean filter: (**a**) original image; (**b**) noisy image; and (**c**) filtered image

SCILAB Code for the Harmonic Mean Filter

```
a = uigetfile("*.*","Select the Image:-");
a = imread(a);
b1 = double(mtlb_double(a));
c = imnoise(a,"gaussian");
d = double(mtlb_double(c));
b = d;
m = (1/9)*ones(3,3);
[r1,c1] = size(mtlb_double(a));
for i = 2:r1-1
 for j = 2:c1-1
  a1 = 1/d(i-1,j-1)+1/d(i-1,j)+1/d(i-1,j+1)+1/d(i,j-1)+1/d(i,j)
+1/d(i,j+1)+1/d(i+1,j-1)+1/d(i+1,j)+1/d(i+1,j+1);
  b(i,j) = 9/a1;
 end;
end;
imshow(b);title("Filtered Image");
```

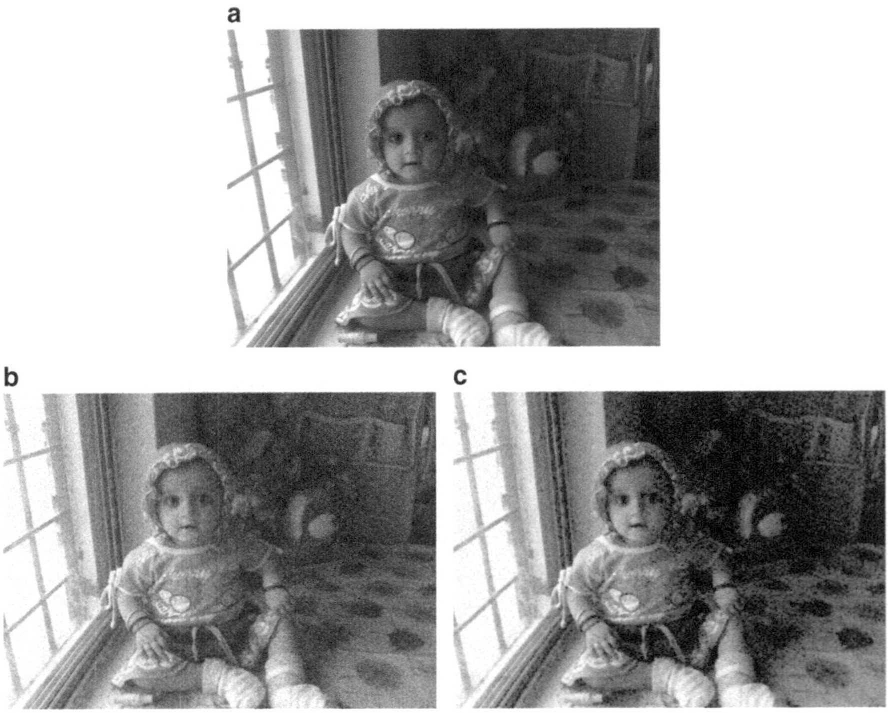

Fig. 4.12 Result from the harmonic mean filter: (**a**) original image; (**b**) noisy image; and (**c**) filtered image

4.5.1.4 Contra Harmonic Mean Filter

The output of the contra harmonic mean filter is given as follows:

$$f'(x,y) = \frac{\sum_{(p,t)\in P_{xy}} g(p,t)^{Q+1}}{\sum_{(p,t)\in P_{xy}} g(p,t)^{Q}} \tag{4.14}$$

where Q is the filter order.

This filter is used to remove the salt and pepper noise from the noisy image. A positive value of Q removes the pepper noise and use of a negative value of Q removes the salt noise. It works as an arithmetic filter when $Q = 0$ and harmonic filter when $Q = -1$. The SCILAB code for this filter is given here and the result from this filter is shown in Fig. 4.13.

SCILAB Code for the Contra Harmonic Mean Filter

```
a = uigetfile("*.*","Select the Image:-");
Q = input("Enter the Order of the filter:");
a = imread(a);
```

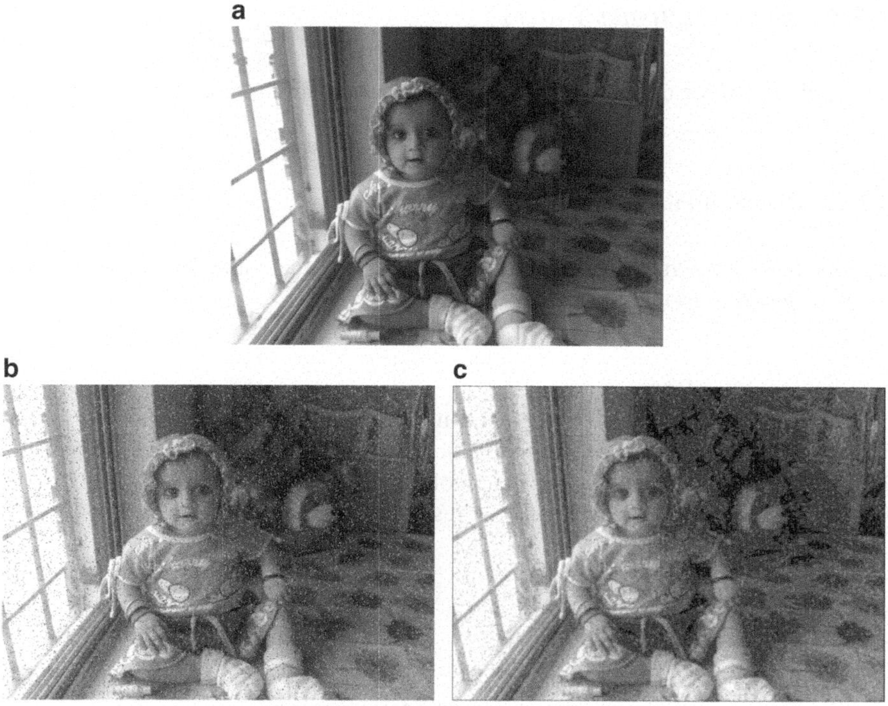

Fig. 4.13 Result from the contra harmonic mean filter: (**a**) original image; (**b**) noisy image; (**c**) filtered image

```
b1 = im2double(a);
c = imnoise(b1,"salt & pepper");
d = c;
b = d;
m = (1/9)*ones(3,3);
[r1,c1] = size(double(a));
for i = 2:r1-1
 for j = 2:c1-1
  a1 = double(d(i-1,j-1))+double(d(i-1,j))+double(d(i-1,j+1))
+mtlb_double(d(i,j-1))+double(d(i,j))+double(d(i,j+1))+double
(d(i+1,j-1))+double(d(i+1,j))+double(d(i+1,j+1));
  a2 = a1 .^mtlb_a(mtlb_double(Q),1);
  a3 = a1 .^mtlb_double(Q);
  b(i,j) = a2 ./a3;
 end;
end;

b = b;
b = b*28;
b=uint8(b);
imshow(b);title("Contra_Harmonic Filtered Image");
```

4.5.2 *Order Statistics Filters*

The response of the order statistics filters is based on ordering or ranking of pixels.

4.5.2.1 Median Filter

The median filter removes a pixel value using the median value of its neighboring pixel. The output of this filter is given as follows:

$$f^{'}(x,y) = \text{median}(p,t) \in P_{xy}(g(p,t)) \tag{4.15}$$

This filter provides excellent results compared with other linear smoothing filters when an image is less corrupted due to blurring. This filter is very useful for removal of unipolar noise (salt or pepper) and bipolar noise (salt and pepper). The SCILAB code for this filter is given here and the result of this filter is shown in Fig. 4.14.

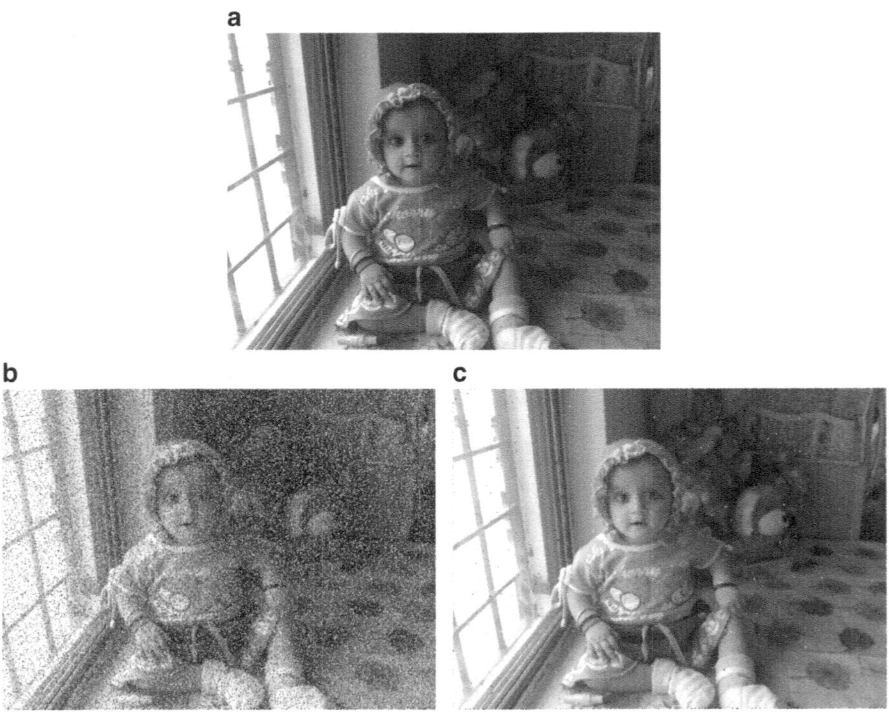

Fig. 4.14 Result from the median filter: (**a**) original image; (**b**) noisy image; and (**c**) filtered Image

SCILAB Code for the Median Filter

```
a = uigetfile("*.*","Select the Image:-");
a = imread(a);
b1 = double(mtlb_double(a));
c = imnoise(a,"salt & pepper",0.2);
d = double(mtlb_double(c));
b = d;
m = (1/9)*ones(3,3);
[r1,c1] = size(mtlb_double(a));
for i = 2:r1-1
 for j = 2:c1-1
  a1 = [d(i-1,j-1),d(i-1,j),d(i-1,j+1),d(i,j-1),d(i,j),d(i,j
+1),d(i+1,j-1),d(i+1,j),d(i+1,j+1)];
  a2 = gsort(a1,"g","i");
  med = a2(5);
  b(i,j) = med;
 end;
end;
b=uint8(b);
imshow(b);title("Median Filtered Image");
```

4.5.2.2 Min Filter

The output of the min filter is given as follows:

$$f^{'}(x,y) = \min(p,t) \in P_{xy}(g(p,t)) \tag{4.16}$$

This filter is used to remove salt noise from the noisy image and to find dark points in the image. The SCILAB code for this filter is given here and the result of this filter is shown in Fig. 4.15.

SCILAB Code for the Min Filter

```
a=uigetfile('*.*','Select the Image:-');
a=imread(a);
b1=double(a);
c=imnoise(a,'salt & pepper',0.2);
d=double(c);
b=d;
m=(1/9)*(ones(3,3));
[r1,c1]=size(a);

for i=2:r1-1
  for j=2:c1-1
    a1=[d(i-1,j-1) d(i-1,j) d(i-1,j+1) d(i,j-1) d(i,j) d(i,j+1)
d(i+1,j-1) d(i+1,j) d(i+1,j+1)];
    a2=gsort(a1);
    med=a2(1);
    b(i,j)=med;
```

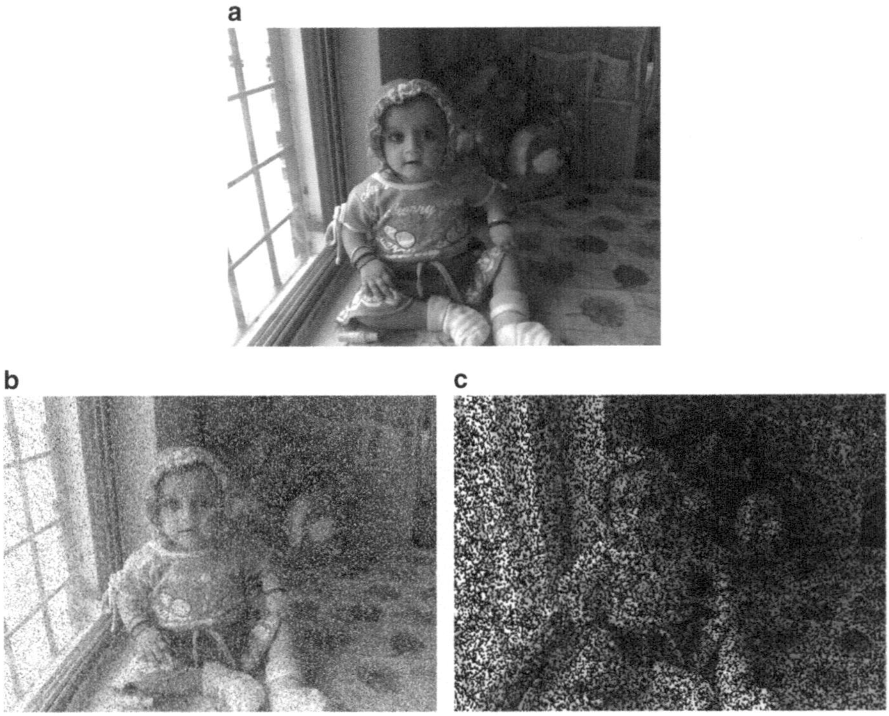

Fig. 4.15 Result from the min filter: (**a**) original image; (**b**) noisy image; and (**c**) filtered image

```
  end
end

b=uint8(b);
imshow(b);title("Min Filtered Image");
```

4.5.2.3 Max Filter

The output of the max filter is given as follows:

$$f^{'}(x,y) = \max(p,t) \in P_{xy}(g(p,t)) \tag{4.17}$$

This filter is used to remove pepper noise from the noisy image and to find bright points in the image. The SCILAB code for this filter is given here and the result of this filter is shown in Fig. 4.16.

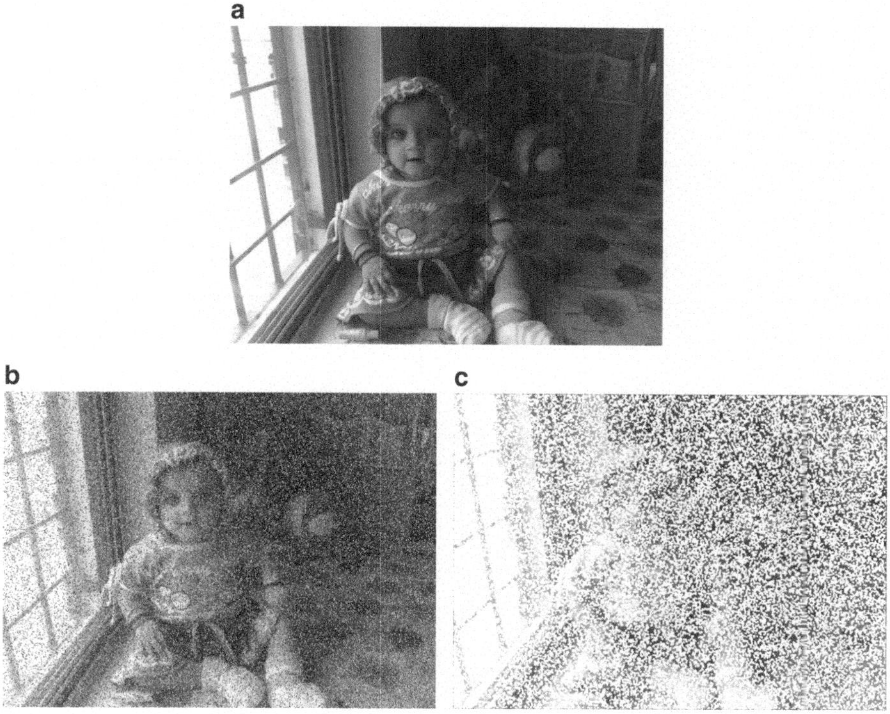

Fig. 4.16 Result from the max filter: (**a**) original image; (**b**) noisy image; and (**c**) filtered image

SCILAB Code for the Max Filter

```
a=uigetfile('*.*','Select the Image:-');
a=imread(a);
b1=double(a);
c=imnoise(a,'salt & pepper',0.2);
d=double(c);
b=d;
m=(1/9)*(ones(3,3));
[r1,c1]=size(a);

for i=2:r1-1
  for j=2:c1-1
    a1=[d(i-1,j-1) d(i-1,j) d(i-1,j+1) d(i,j-1) d(i,j) d(i,j+1) d
(i+1,j-1) d(i+1,j) d(i+1,j+1)];
    a2=gsort(a1);
    med=a2(1);
    b(i,j)=med;
  end
end

b=uint8(b);
imshow(b);title("Min Filtered Image");
```

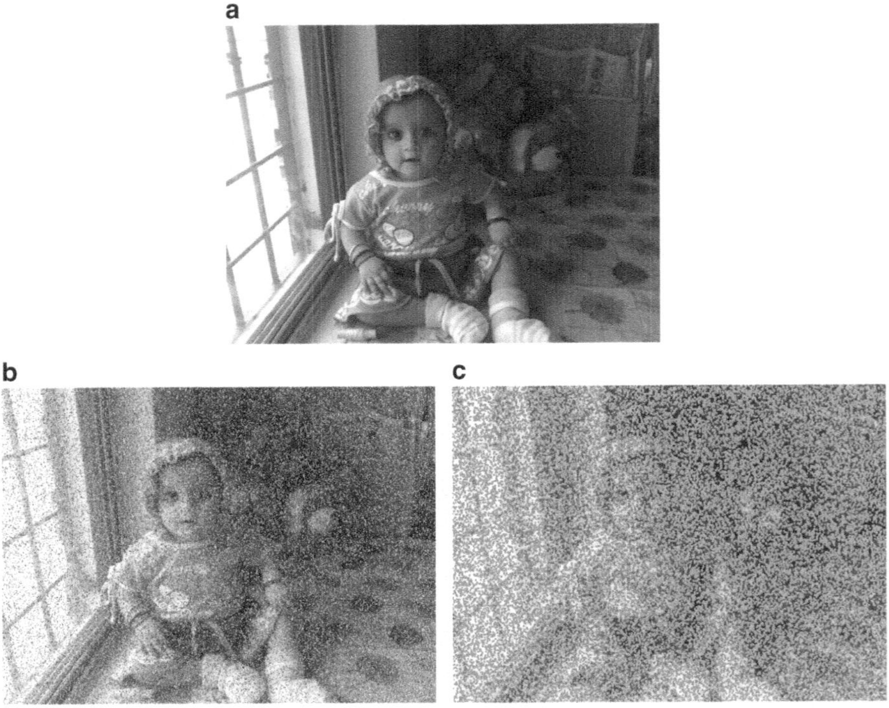

Fig. 4.17 Result from the mid-point filter: (**a**) original image; (**b**) noisy image; and (**c**) filtered image

4.5.2.4 Mid-Point Filter

The output of the mid-point filter is given as follows:

$$f'(x,y) = \frac{1}{2}\left(\max(p,t) \in P_{xy}\left(g(p,t)\right) + \min(p,t) \in P_{xy}\left(g(p,t)\right)\right) \tag{4.18}$$

The midpoint between the minimum and maximum values of intensity pixels is calculated. This filter is a combination of an order statistics filter and average filter. It removes Gaussian noise and uniform noise from the noisy image. The SCILAB code for this filter is given here and the result of this filter is shown in Fig. 4.17.

SCILAB Code for the Mid-Point Filter

```
a=uigetfile('*.*','Select the Image:-');
a=imread(a);
b1=double(a);
c=imnoise(a,'salt & pepper',0.2);
d=double(c);
b=d;
```

```
m=(1/9)*(ones(3,3));
[r1,c1]=size(a);

for i=2:r1-1
  for j=2:c1-1
    a1=[d(i-1,j-1) d(i-1,j) d(i-1,j+1) d(i,j-1) d(i,j) d(i,j-1) d
(i+1,j-1) d(i+1,j) d(i+1,j+1)];
    a2=gsort(a1);
    med=a2(9);
    med1=a2(1);
    med=(1/2)*(med+med1);
    b(i,j)=med;
  end
end
b=uint8(b);
imshow(b);title("Mid-Point Filtered Image");
```

4.5.2.5 Alpha Trimmed Mean Filter

An alpha trimmed mean filter occurs when the highest and lowest pixel values from $d/2$ are deleted from the noisy image, the resultant image being denoted $g_r(x, y)$, and the filter finds the averages values of the pixels in the image. The output of this filter is given as follows:

$$f'(x,y) = \frac{1}{MN-d}\sum\nolimits_{(p,t)\in P_{xy}} g_r(x,y) \tag{4.19}$$

where d is the distance from 0 to $M*N-1$.

If value of $d = 0$, then this filter works as an arithmetic mean filter, and when $d = 1$ this filter works as a mean filter. This filter removes Gaussian noise and salt and pepper noise from the noisy image (Fig. 4.18).

SCILAB Code for the Alpha Trimmed Mean Filter

```
z=uigetfile('*.*','Select an image');
z=imread(z);
a=double(imnoise(z,'gaussian'));
r=size(a,1);
c=size(a,2);
d=input('Enter the value of d(0 to 8): ');
for i=6-(d/2):r-(1+(d/2))
  for j=6-(d/2):c-(1+(d/2))
    n=a(i-1,j-1)+a(i-1,j)+a(i-1,j+1)+a(i,j-1)+a(i,j)+a(i,j+1)
+a(i+1,j-1)+a(i+1,j)+a(i+1,j+1);
    new(i,j)=(1/(9-(d/2)))*n;
  end
end
new=uint8(new);
imshow(new);title("Alpha Trimmed Mean Filtered Image");
```

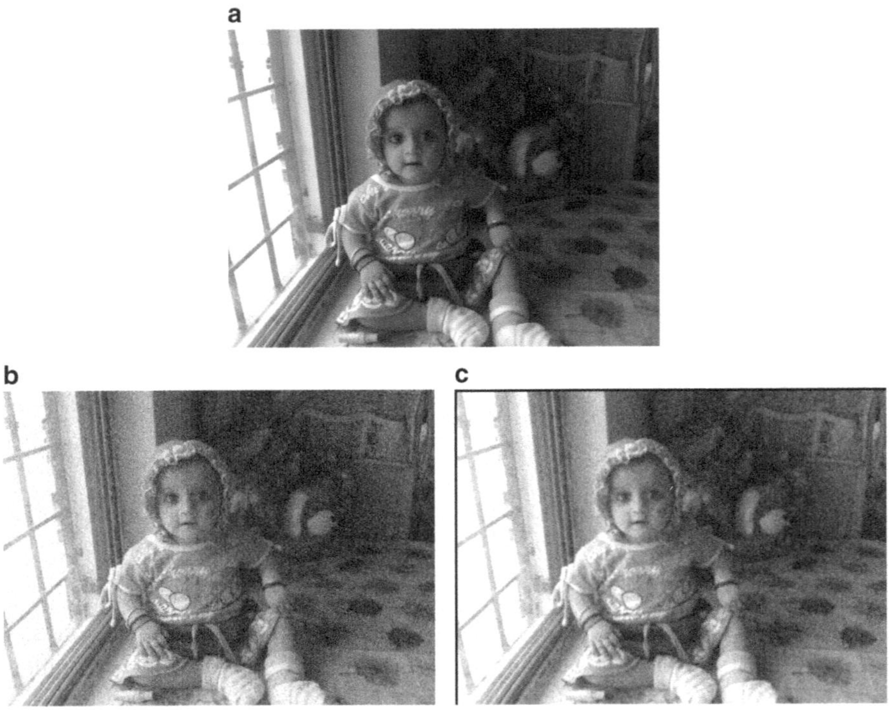

Fig. 4.18 Result from the alpha trimmed mean filter: (**a**) original image; (**b**) noisy image; and (**c**) filtered image

4.5.3 Adaptive Filters

Adaptive filters change the statistical characteristics of an image inside the filter region P_{xy}. This filter provides the same performance for noise removal but adds less blurring than a mean filter. This provides a better result than the other two types of filters. These filters are divided into two types: the adaptive local noise reduction filter and adaptive median filter. The algorithms for these filters were beautifully described by Gonzales and Woods in 2008 [1, 2]. The adaptive local noise reduction filter is used for estimation of noise presented in a selected area of the noisy image and filters noise presented in the selected area. The adaptive median filter is used to remove impulse noise which can't be removed by a traditional median filter. The SCILAB code for the adaptive median filter is given here and the result of this filter is shown in Fig. 4.19.

SCILAB Code for the Adaptive Median Filter

```
A = uigetfile('*.*','Select an image');
B = imread(A);
sz = size(double(B),1)*size(double(B),2);
```

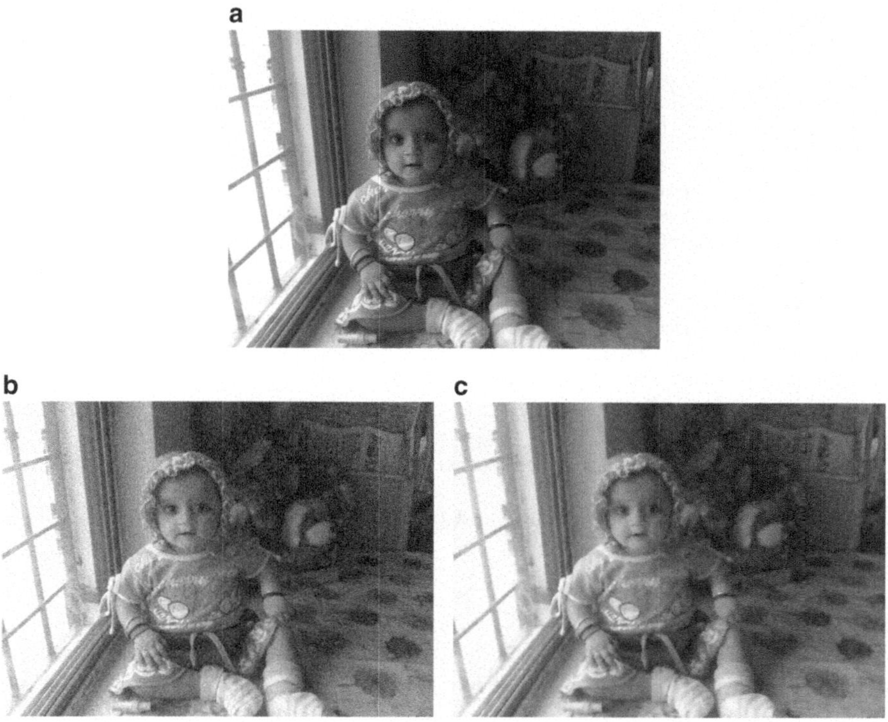

Fig. 4.19 Result of the adaptive median filter: (**a**) original image; (**b**) noisy image; and (**c**) filtered image

```
B = imnoise(B,"gaussian",0,0.005);
B = double(B);
//Define the window size mxn
M = 5;
N = 5;
//Pad the matrix with zeros on all sides
Z= [floor(M/2),floor(N/2)]
C =zeros(B,Z);
v0 = [size(B,1),size(B,2)];lvar = zeros(v0(1),v0(2));
v0 = [size(B,1),size(B,2)];lmean = zeros(v0(1),v0(2));
v0 = [size(B,1),size(B,2)];temp = zeros(v0(1),v0(2));
v0 = [size(B,1),size(B,2)];NewImg = zeros(v0(1),v0(2));

for i = 1:size(mtlb_double(C),1)-(M-1)
 for j = 1:size(mtlb_double(C),2)-(N-1)

  temp = C(i:i+(M-1),j:j+(N-1));
  tmp = mtlb_e(temp,":");
  lmean(i,j) = mean(double(tmp),"m");
  lvar(i,j) = mtlb_s(mean(mtlb_double(tmp).^2,"m"),mean
(mtlb_double(tmp),"m").^2);
```

```
 end;
end;

//Noise variance and average of the local variance
nvar = sum(lvar(:),firstnonsingleton(lvar(:)))/sz;

//If noise_variance > local_variance then
local_variance=noise_variance
lvar = mtlb_max(lvar,nvar);

//Final_Image = B- (noise variance/local variance)*
(B-local_mean);
NewImg = nvar ./lvar;
NewImg = NewImg .*mtlb_s(B,lmean);
NewImg = mtlb_s(B,NewImg);

//Convert the image to uint8 format.
NewImg = NewImg;
imshow(NewImg);title("Restored Image using adaptive median
filter");
```

4.6 Wiener Filtering

An image can restored from a degraded image using the simple method of forward Fourier transform, which is estimated in the following form:

$$\widehat{F}(u,v) = \frac{G(u,v)}{H(u,v)} \tag{4.20}$$

The corresponding estimate pixel values of the image are then obtained by taking the inverse Fourier transform. This approach is called inverse filtering. This is the basic restoration process for images using deconvolution when the image is degraded by a low-pass filter as this form of filtering is very sensitive to additive noise. This process reduces one degraded pixel in an image at a time to develop a restoration algorithm for each possible degraded pixel and combines the results of each process.

However, inverse filtering process is a time-consuming process. Wiener filtering provides an optimal solution between inverse filtering and noise smoothing [3, 4]. It removes the additive noise and inverts the blurring simultaneously. The SCILAB code for the Wiener filter is given here and result of this filter is shown in Fig. 4.20.

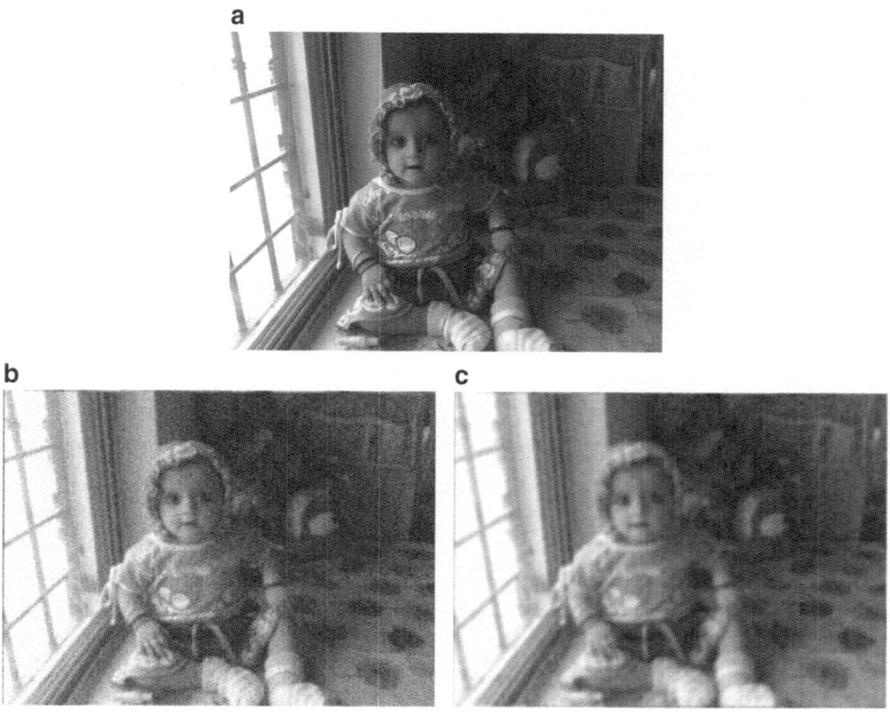

Fig. 4.20 Result from the Wiener filter: (**a**) original image; (**b**) noisy image; and (**c**) filtered image

SCILAB Code for Wiener Filter

```
x = uigetfile('*.*','Select an image');
x=double(imread(x));
sigma = 50;
gamma = 1;
alpha = 1;
[M N]=size(x);
h = ones(5,5)/25;
Freqa = fft2(x);
Freqh = fft2(h,M,N);
y = real(ifft2(Freqh.*Freqa))+25*randn(M,N);
Freqy = fft2(y);
Powy = abs(Freqy).^2/(M*N);
sFreqh = Freqh.*(abs(Freqh)>0)+1/gamma*(abs(Freqh)==0);
iFreqh = 1./sFreqh;
iFreqh = iFreqh.*(abs(Freqh)*gamma>1)...
  +gamma*abs(sFreqh).*iFreqh.*(abs(sFreqh)*gamma<=1);
Powy = Powy.*(Powy>sigma^2)+sigma^2*(Powy<=sigma^2);
Freqg = iFreqh.*(Powy-sigma^2)./(Powy-(1-alpha)*sigma^2);
ResFreqa = Freqg.*Freqy;
Resa = real(ifft2(ResFreqa));
imshow(uint8(Resa)),title('Weiner Filtered Image');
```

Bibliography

1. Gonzalez, R., & Woods, R. (2008). *Digital image processing*. Upper Saddle River: Pearson Education India.
2. Gonzalez, R., Woods, R., & Eddins, L. (2009). *Digital image processing using MATLAB*. New Delhi: TATA McGraw-Hill Education.
3. Wiener, N. (1942). The interpolation, extrapolation and smoothing of stationary time series. Report of the Services 19, Research Project DIC-6037 MIT.
4. Baraniuk, R. (2005). *Digital image processing*. Houston: Rice University.

Chapter 5
Morphological Image Processing

5.1 Introduction

Morphological image processing (or morphology) describes various image processing operations which deal with the shape of features in an image [1–4]. These operations remove imperfections in the image and are applied on binary images. The shape features of the image (i.e., edges, holes, corners, cracks) can be extracted in this process using various shaped structuring elements. This process is used in many industrial computer vision applications such as recognition of objects, segmentation of images, and finding defects.

5.1.1 Structuring Elements: Hits and Fits

Structuring elements are a small binary images with value of 0 or 1. Any element with the below characteristics are considered to be structuring elements:

- The dimension of the elements is defined by a specified matrix size.
- The pattern of ones and zeros specifies the structure element shape.
- The origin of the structuring element is usually one of its pixels, although, generally, the origin can be outside the structuring element. A few examples of structuring elements are shown in Fig. 5.1.

In common practice, structuring elements are chosen as a matrix with an odd dimension size and the origin of the elements defined by the center of the matrix. The role of structuring elements in morphological image processing is the same as the role of convolution kernels in spatial linear filtering.

Fit All *on pixels* in the structuring element cover *on pixels* in the image (as shown in Fig. 5.1a).

R. M. Thanki, A. M. Kothari, *Digital Image Processing using SCILAB*,
https://doi.org/10.1007/978-3-319-89533-8_5

a

1	1	1	1	1
1	1	1	1	1
1	1	1	1	1
1	1	1	1	1
1	1	1	1	1

b

		1		
	1	1	1	
1	1	1	1	1
	1	1	1	
		1		

c

1	1	1
1	1	1
1	1	1

Fig. 5.1 A few examples of structuring elements. (**a**) 5 * 5 square structuring element (**b**) diamond structuring element (**c**) 3 * 3 square structuring element

Hit Any *on pixel* in the structuring element covers an *on pixel* in the image (as shown in Fig. 5.1b).

All operations related to morphological processing depend on the structuring elements, fit and hit. The basic morphological processing is similar to spatial filtering. In this process, the structuring element is moved across every pixel in the original image to give a new pixel value for the processed image. The new value of the pixel depends on the operation performed on it.

5.2 Fundamental Morphology Operations

There are two fundamental morphological operations: erosion and dilation. Detail and implementation of these operations are given in the following sections.

5.2.1 Erosion

Erosion of image f by structuring element s of size $M \times N$ is given by $f \ominus s$. The structuring element s is positioned with its origin at (x, y) and the new value of the pixel is calculated using the following rule:

$$g(x,y) = 1 \text{ if } s\,\textbf{fits}\,f \tag{5.1}$$

$$g(x,y) = 0, \text{ otherwise} \tag{5.2}$$

Or it can also be given as:

$$g(x,y) = \min(f(x-i,\ \ y-j),\ \ s(i,j)),\ \ 0 \le i \le M-1,\ \ 0 \le j \le N-1 \tag{5.3}$$

If the structuring element is shown in Fig. 5.2a and the original binary image is shown in Fig. 5.2b, c, then, according to Eq. 5.3, the processed binary image by the erosion operation is given in Fig. 5.3.

The SCILAB code for the erosion operation is given here and the result of this operation is given in Fig. 5.4.

Fig. 5.2 (**a**) Structuring element. (**b**) Binary values of image. (**c**) Original image

a

1	1	1
1	1	1
1	1	1

b

0	0	0	0	0	0	0	0
0	0	0	1	1	1	0	0
0	0	1	1	1	1	0	0
0	1	1	1	1	0	0	0
0	1	1	1	1	0	0	0
0	1	1	1	1	0	0	0
0	0	0	0	0	0	0	0
0	0	0	0	0	0	0	0

c

Fig. 5.3 (**a**) Processed binary values of image. (**b**) Processed image after erosion operation

a

0	0	0	0	0	0	0	0
0	0	0	0	0	0	0	0
0	0	0	0	0	0	0	0
0	0	0	0	0	0	0	0
0	0	1	0	0	0	0	0
0	1	1	0	0	0	0	0
0	0	0	0	0	0	0	0
0	0	0	0	0	0	0	0

b

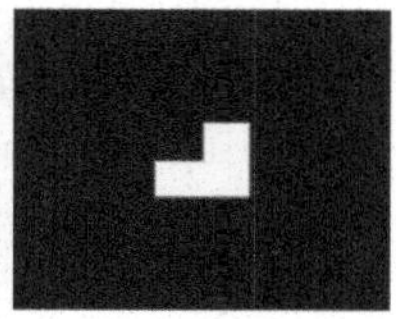

Fig. 5.4 Result from the erosion operation: (**a**) original image; and (**b**) processed image

SCILAB Code for Erosion

```
a=uigetfile('*.*','Select the Image');
a=imread(a);
d=a;
[r,c]=size(d);
m=[1 1 1;1 1 1;1 1 1];
// m=ones(5,5);
for i=2:1:r-1
  for j=2:1:c-1

    new=[(m(1)*d(i-1,j-1)) (m(2)*d(i-1,j)) (m(3)*d(i-1,j
+1)).......
       (m(4)*d(i,j-1)) (m(5)*d(i,j)) (m(6)*d(i,j+1)).......
       (m(7)*d(i+1,j-1)) (m(8)*d(i+1,j)) (m(9)*d(i+1,j+1))];
    A1(i,j)=min(new);
  end
end
imshow(A1);title('Processed Image - Erosion');
```

Observations from the erosion process:

- Erosion can repair breaks in black objects and can split apart white objects.
- Erosion can repair intrusions in black objects and can strip away extrusions in white objects.
- Erosion enlarges black and shrinks white objects.

5.2.2 *Dilation*

The dilation of image f by structuring element s of size $M \times N$ is given by $f \oplus s$. The structuring element s is positioned with its origin at (x, y) and the new value of the pixel is calculated using the following rule:

$$g(x,y) = 1 \text{ if } s\,\textbf{hits}\,f \tag{5.4}$$
$$g(x,y) = 0, \text{otherwise} \tag{5.5}$$

Or it can also be given as:

$$g(x,y) = \max\big(f(x-i, y-j), s(i,j)\big), \quad 0 \le i \le M-1, \quad 0 \le j \le N-1 \tag{5.6}$$

If the structuring element is shown in Fig. 5.2a and the original binary image is shown in Fig. 5.2b, c, then, according to Eq. 5.3, the processed binary image by the dilation operation is given in Fig. 5.5.

The SCILAB code for the dilation operation is given here and the result of this operation is given in Fig. 5.6.

SCILAB Code for Dilation

```
a=uigetfile('*.*','Select the Image');
a=imread(a);
d=a;
[r,c]=size(d);
m=[1 1 1;1 1 1;1 1 1];
```

a

0	0	0	0	0	0	0	0
0	0	1	1	1	1	1	0
0	1	1	1	1	1	1	0
0	1	1	1	1	1	1	0
0	1	1	1	1	1	1	0
0	1	1	1	1	1	1	0
0	1	1	1	1	1	1	0
0	0	0	0	0	0	0	0

b

Fig. 5.5 (**a**) Processed binary values of image. (**b**) Processed image after the dilation operation

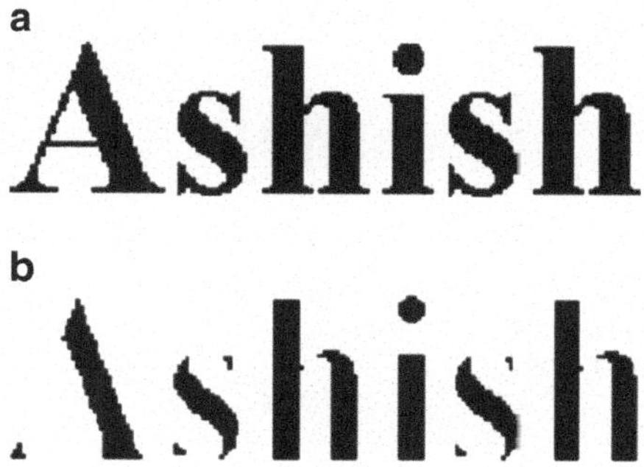

Fig. 5.6 Result from the dilation operation: (**a**) original image; and (**b**) processed image

```
//m=ones(5,5);
for i=2:1:r-1
  for j=2:1:c-1

    new=[(m(1)*d(i-1,j-1)) (m(2)*d(i-1,j)) (m(3)*d(i-1,j+1))
.......
         (m(4)*d(i,j-1)) (m(5)*d(i,j)) (m(6)*d(i,j+1)).......
         (m(7)*d(i+1,j-1)) (m(8)*d(i+1,j)) (m(9)*d(i+1,j+1))];
         A1(i,j)=max(new);
  end
end
imshow(A1);title('Processed Image - Dilation');
```

Observations from the dilation process:

- Dilation can repair breaks in white objects and can split apart black objects.
- Dilation can repair intrusions in white objects and can strip away extrusions in black objects.
- Dilation enlarges white and shrinks black objects.

5.3 Compound Morphology Operations

Compound morphology operations are more interesting morphological operations that combine the erosion operation and dilation operation. These compound operations are known as opening and closing.

5.3.1 Opening

The opening of image f by structuring elements s denoted as $f \circ s$. This operation can be obtained using the erosion process followed by the dilation process. The opening operation for the image is shown in Fig. 5.7.

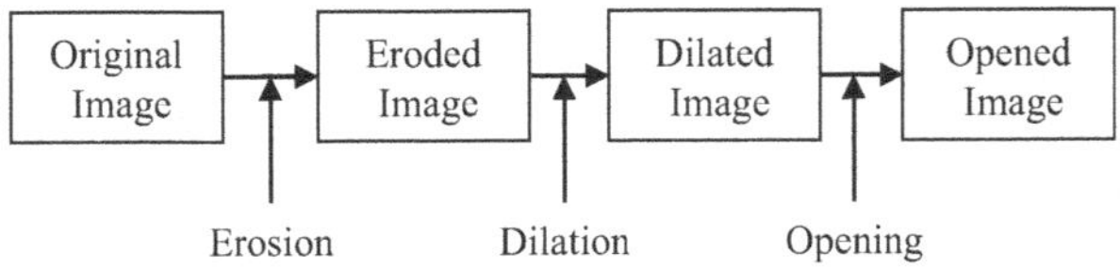

Fig. 5.7 Opening operation for the image

a

1	1	1	0	0	0
1	1	1	0	0	0
1	1	1	1	0	0
0	0	1	1	1	1
0	0	0	1	1	1
0	0	0	1	1	1

b

Fig. 5.8 Image for opening operation and closing operation

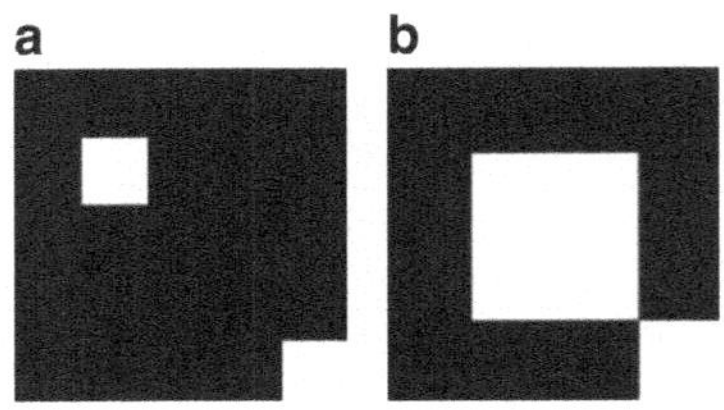

Fig. 5.9 Processed imagine after (**a**) erosion operation; and (**b**) opening operation (dilation of output of erosion operation)

Fig. 5.10 (**a**) Original image. (**b**) Processed image after erosion. (**c**) Processed image after dilation—opening

Mathematically, the opening operation is given as follows:

$$f \circ s = (f \ominus s) \oplus \mathrm{s} \tag{5.7}$$

If the structuring element is shown in Fig. 5.2a and the original binary image is shown in Fig. 5.8a, b, then, according to Eq. 5.7, the processed binary image using the opening operation is given in Fig. 5.8.

The SCILAB code for the opening operation is given here and the result of this operation is given in Fig. 5.10.

SCILAB Code for Opening

```
d=[zeros(1,9);zeros(1,9);0 0 0 1 1 1 0 0 0;0 0 1 1 1 1 0 0 0;0 0 1 1 1 1 0
0 0;0 0 1 1 1 1 0 0 0;zeros(1,9);zeros(1,9);zeros(1,9);]
a=uigetfile('*.*','Select the Image');
a=imread(a);
d=a;
d1=d;
[r,c]=size(d);
m=[1 1 1;1 1 1;1 1 1];
for i=2:1:r-1
  for j=2:1:c-1

    new=[(m(1)*d(i-1,j-1)) (m(2)*d(i-1,j)) (m(3)*d(i-1,j+1))
.......
      (m(4)*d(i,j-1)) (m(5)*d(i,j)) (m(6)*d(i,j+1)).......
      (m(7)*d(i+1,j-1)) (m(8)*d(i+1,j)) (m(9)*d(i+1,j+1))];
      A2(i,j)=min(new);
  end
end

d = A2;
[r,c]=size(d);
for i=2:1:r-1
  for j=2:1:c-1

    new=[(m(1)*d(i-1,j-1)) (m(2)*d(i-1,j)) (m(3)*d(i-1,j+1))
.......
      (m(4)*d(i,j-1)) (m(5)*d(i,j)) (m(6)*d(i,j+1)).......
      (m(7)*d(i+1,j-1)) (m(8)*d(i+1,j)) (m(9)*d(i+1,j+1))];
      A1(i,j)=max(new);
  end
end

imshow(A1);title('Dilation - Opening');
```

Observations from the opening process:

- Opening smoothes the contours of the image.
- Opening breaks down narrow bridges.
- Opening eliminates protrusions.
- Thus, opening isolates objects which may be just touching one another.
- Opening is used in the analysis of wear particles in engine oil, ink particles in recycled paper and, most importantly, in the study of cells in cytology.

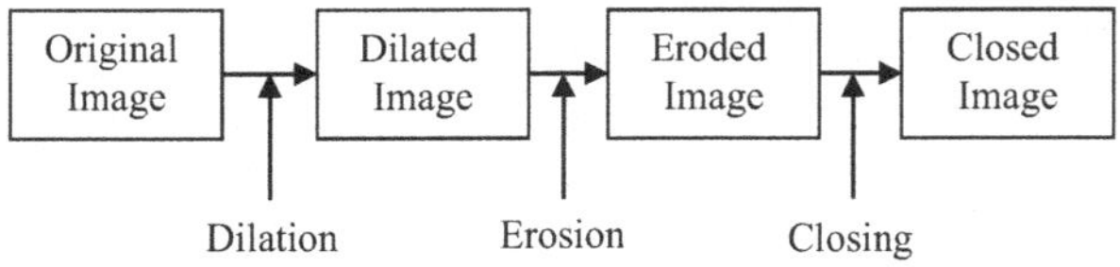

Fig. 5.11 Closing operation for image

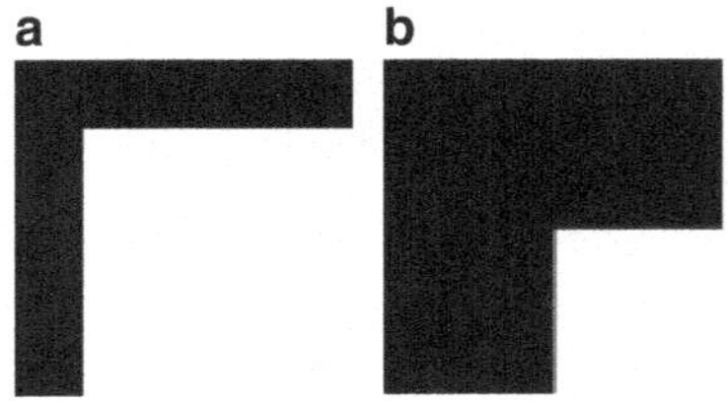

Fig. 5.12 Processed image after (**a**) dilation operation; and (**b**) closing operation (erosion of output of dilation operation)

Fig. 5.13 (**a**) Original image. (**b**) Processed image after dilation. (**c**) Processed image after erosion—closing

5.3.2 *Closing*

The closing of image f by structuring elements s is denoted as $f \bullet s$. This operation can be obtained using the dilation process followed by the erosion process. The closing operation for an image is shown in Fig. 5.11.

Mathematically, the opening operation is given as follows:

$$f \bullet s = (f \oplus s) \ominus s \tag{5.8}$$

If the structuring element is shown in Fig. 5.2a and the original binary image is shown in Fig. 5.8a, b, then, according to Eq. 5.8, the processed binary image by the closing operation is given in Fig. 5.11.

The SCILAB code for the closing operation is given here and the result of this operation is given in Fig. 5.13.

SCILAB Code for Closing

```
d=[zeros(1,9);zeros(1,9);0 0 0 1 1 1 0 0 0;0 0 1 1 1 1 0 0 0;0 0 1 1 1 1 0
0 0;0 0 1 1 1 1 0 0 0;zeros(1,9);zeros(1,9);zeros(1,9);]
a=uigetfile('*.*','Select the Image');
a=imread(a);
d=a;
d1=d;
[r,c]=size(d);
m=[1 1 1;1 1 1;1 1 1];
for i=2:1:r-1
  for j=2:1:c-1

    new=[(m(1)*d(i-1,j-1)) (m(2)*d(i-1,j)) (m(3)*d(i-1,j+1))
.......
      (m(4)*d(i,j-1)) (m(5)*d(i,j)) (m(6)*d(i,j+1)).......
      (m(7)*d(i+1,j-1)) (m(8)*d(i+1,j)) (m(9)*d(i+1,j+1))];
      A2(i,j)=max(new);
  end
end

d = A2;
[r,c]=size(d);
for i=2:1:r-1
  for j=2:1:c-1

    new=[(m(1)*d(i-1,j-1)) (m(2)*d(i-1,j)) (m(3)*d(i-1,j+1))
.......
      (m(4)*d(i,j-1)) (m(5)*d(i,j)) (m(6)*d(i,j+1)).......
      (m(7)*d(i+1,j-1)) (m(8)*d(i+1,j)) (m(9)*d(i+1,j+1))];
      A1(i,j)=min(new);
  end
end

imshow(A1);title('Erosion - Closing');
```

Observations from the closing process:

- Closing tends to fuse narrow breaks.
- Closing eliminates small holes.
- Closing is used to simplify the process of assessing the separation of particles.

5.4 Hit or Miss Transform

Hit or miss transform is a powerful approach for finding objects and their locations in an image. It can be defined as combination erosion and dilation. It is used to find specific shapes such as squares, triangles, or cornets in the image. In the erosion process, the 0 bits in the structuring element work as "don't care" conditions and they don't really require that the image be on or off at that point. Thus, only pixels

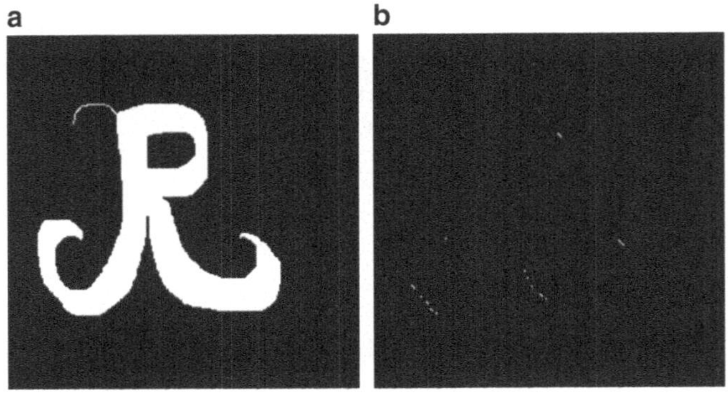

Fig. 5.14 Result of hit or miss transform: (**a**) original image; and (**b**) processed image

with 1 bit are presented inside the object. The combination of the erosion process operation and dilation process operator produces the "hit and miss" operator. This operation has two structuring elements with one element for the hit operation and the second element for the miss operation. This operation can be defined as follows: if the structuring elements *s1* (hit) and *s2* (miss) are applied to the image *f*, then:

$$f \otimes S = (f \ominus s1) \cap (f^C \ominus s2) \tag{5.9}$$

where $S = (s1, s2)$

This gives us a form of binary template matching. For example, the following structuring elements give an "upper right" corner detector:

S1 =

0	0	0
1	1	0
0	1	0

S2=

0	1	1
0	0	1
0	0	0

The *s1* element finds the points with connected left and lower neighbors, and the *s2* element finds the points without upper, upper right, and right neighbors. The SCILAB code for the hit or miss transform is given here and the result of this operation is given in Fig. 5.14.

SCILAB Code for the Hit and Miss Transform

```
z=uigetfile('*.*','Select an image file');
a=imread(z);

j=[0 0 0 ; 255 255 0 ; 0 255 0];
k=[0 255 255 ; 0 0 255 ; 0 0 0];

r=size(a,1);
c=size(a,2);
```

```
for i=2:r-1
  for j=2:c-1
   if (a(i-1,j-1)==j(1) &a(i-1,j)==j(2) &a(i-1,j+1)==j(3) &a(i,
j-1)==j(4)&a(i,j)==j(5) &a(i,j+1)==j(6) &a(i+1,j-1)==j(7) &a
(i+1,j)==j(8) &a(i+1,j+1)==j(9))
      new(i,j)=255;
    else
      new(i,j)=0;
    end
  end
end

for i=1:r
  for j=1:c
    aa(i,j)=255-a(i,j);
  end
end

for i=2:r-1
  for j=2:c-1
    if (aa(i-1,j-1)==k(1) &aa(i-1,j)==k(2) &aa(i-1,j+1)==k(3) &
aa(i,j-1)==k(4) &aa(i,j)==k(5) &aa(i,j+1)==k(6) &aa(i+1,j-1)
==k(7) &aa(i+1,j)==k(8) &aa(i+1,j+1)==k(9))
      new1(i,j)=255;
    else
      new1(i,j)=0;
    end
  end
end

r=size(new,1);
c=size(new,2);

for i=2:r-1
  for j=2:c-1
    new2(i,j)=new(i,j)+new1(i,j);
  end
end

imshow(new2);title('Hit Miss Transformed Image');
```

5.5 Some Morphological Algorithms

We have already studied and demonstrated fundamental as well as compound morphological operations. Using the concept of these operations, in the following sections we now provide SCILAB codes for some basic morphological algorithms to extract object features, describe shapes, and recognize patterns.

a

0	0	0	0	0	0	0	0	0
0	0	0	0	0	0	0	0	0
0	0	0	1	1	1	0	0	0
0	0	1	0	0	1	0	0	0
0	0	1	0	0	1	0	0	0
0	0	1	1	1	1	0	0	0
0	0	0	0	0	0	0	0	0
0	0	0	0	0	0	0	0	0

b

Fig. 5.15 Processed images after application of the boundary extraction operation

Fig. 5.16 Result of the boundary extraction operation: (**a**) original image; and (**b**) processed image

5.5.1 Boundary Extraction

In a binary image, an object set with a pixel value of 1 bit is a connected component. The boundary pixels of those object pixels are those whose neighbor pixels have at least a value of 0 bits. Boundary extraction of an image f first erodes f using a structuring element s and then takes the set difference between f and its erosion. The structuring element for this operation must be isotropic; ideally, a circle. But in image processing, a 3×3 matrix of ones is often used. Therefore, the boundary of a set f is obtained as follows:

$$\text{Internal Boundary } (f) = f - (f \ominus s) \tag{5.10}$$

$$\text{External Boundary } (f) = (f \oplus s) - f \tag{5.11}$$

If the structuring element is shown in Fig. 5.2a and the original binary image is shown in Fig. 5.2b, c, then, according to Eqs 5.10 and 5.11, the processed binary image by the boundary extraction operation is given in Fig. 5.15.

The SCILAB code for boundary extraction is given here and the result of this operation is given in Fig. 5.16.

SCILAB Code for Boundary Extraction

```
a=uigetfile('*.*','Select the Image');
a=imread(a);

d=a;
[r,c]=size(d);
m=[1 1 1;1 1 1;1 1 1];
for i=2:1:r-1
  for j=2:1:c-1

    new=[(m(1)*d(i-1,j-1)) (m(2)*d(i-1,j)) (m(3)*d(i-1,j+1))
.......
      (m(4)*d(i,j-1)) (m(5)*d(i,j)) (m(6)*d(i,j+1)).......
      (m(7)*d(i+1,j-1)) (m(8)*d(i+1,j)) (m(9)*d(i+1,j+1))];
      A2(i,j)=min(new);
      aa(i,j)=d(i,j)-A2(i,j);
  end
end
imshow(aa);title('Boundary Extracted Image');
```

5.5.2 *Thinning*

The thinning operation is similar to erosion, but it doesn't cause the disappearance of object components. It is a fundamental step for structural representation of the pattern shape. Thinning can be used in various applications such as industrial part inspection, fingerprint recognition, optical character recognition, and biomedical diagnosis. Mathematically, it is given as follows:

$$f \times s = f - (f \otimes s) = f \cap (f \otimes s)^{C} \quad (5.13)$$

These processes are based on a search and delete process. The thinning process removes pixels form the outside edges of an object in the image.

The SCILAB code for thinning is given here and result of this operation is given in Fig. 5.17.

SCILAB Code for Thinning

```
z=uigetfile('*.*','Select an image file');
a=imread(z);

s=[255 255 255 ; 255 255 255 ; 255 255 255];

r=size(a,1);
c=size(a,2);
```

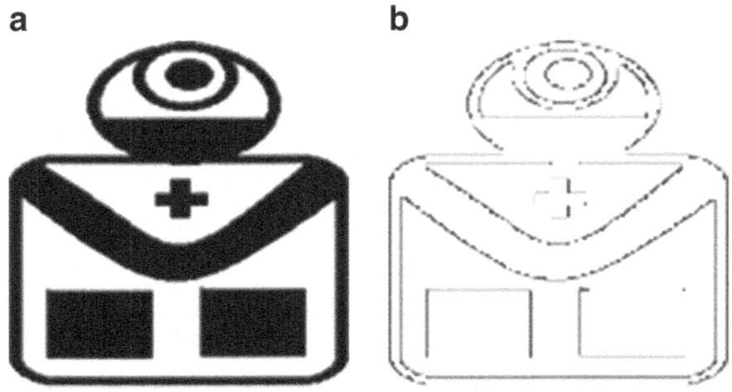

Fig. 5.17 Result of the thinning operation: (**a**) original image; and (**b**) processed image

```
for i=2:r-1
  for j=2:c-1
   if (a(i-1,j-1)==s(1) | a(i-1,j)==s(2) | a(i-1,j+1)==s(3) | a(i,
j-1)==s(4) | a(i,j)==s(5) | a(i,j+1)==s(6) | a(i+1,j-1)==s(7) | a
(i+1,j)==s(8) | a(i+1,j+1)==s(9))
      new(i,j)=255;
    else
      new(i,j)=0;
    end
    new1(i,j)=255-new(i,j);
    new2(i,j)=a(i,j)+new1(i,j);
  end
end

imshow(new2);title('Thinning Image');
```

5.5.3 *Thickening*

The thickening operation is similar to dilation, but it does not cause merging of disconnected objects. It is used to grow a portion of an object and can be represented in terms of the hit or miss operation and union as follows:

$$f^{*'}s = f \cup (f \otimes s) \tag{5.14}$$

The SCILAB code for thickening is given here and result of this operation is given in Fig. 5.18.

SCILAB Code for Thickening

```
z=uigetfile('*.*','Select an image file');
a=imread(z);

s=[255 255 255 ; 255 255 255 ; 255 255 255];
```

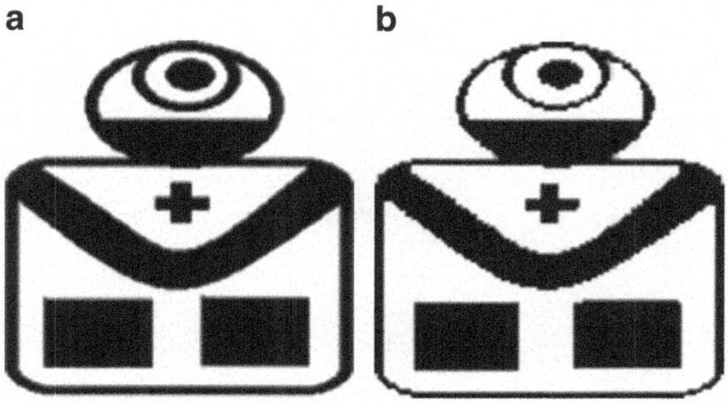

Fig. 5.18 Result of the thickening operation: (**a**) original image; and (**b**) processed image

```
r=size(a,1);
c=size(a,2);

for i=2:r-1
  for j=2:c-1
   if (a(i-1,j-1)==s(1) | a(i-1,j)==s(2) | a(i-1,j+1)==s(3) | a(i,
j-1)==s(4) | a(i,j)==s(5) | a(i,j+1)==s(6) | a(i+1,j-1)==s(7) | a
(i+1,j)==s(8) | a(i+1,j+1)==s(9))
      new(i,j)=255;
    else
      new(i,j)=0;
    end
    new1(i,j)=a(i,j)+new(i,j);
  end
end
imshow(new1);title('Thickening Image');
```

Bibliography

1. Gonzalez, R., & Woods, R. (2008). *Digital image processing*. Upper Saddle River: Pearson Education India.
2. Gonzalez, R., Woods, R., & Eddins, L. (2009). *Digital image processing using MATLAB*. New Delhi: TATA McGraw-Hill Education.
3. Morphological Image Processing, website: https://www.cs.auckland.ac.nz/courses/compsci773s1c/lectures/ImageProcessing-html/topic4.htm, last access: January, 2018.
4. Efford, N. (2000). *Digital image processing: A practical introduction using JAVA*. New Delhi: Pearson Education.

Chapter 6
Image Segmentation

6.1 Introduction

Image segmentation is used to separate an image into constituent parts based on some image attributes. This process reduces a huge amount of unnecessary data, retaining only important data for image analysis. It also converts bitmap data into better structured data that are easier to interpret. Image segmentation can be performed in two ways: using similarity properties of pixels and discontinuity properties of pixels. The similarity properties of pixels means that targeted pixels have the same gray level intensity, while discontinuity properties of pixels means that boundary pixels have a different gray level intensity than groups of targeted pixels. The image segmentation process mainly extracts three features: lines, points, and edges in the image. The process of detection of these features is similar to mask processing. The following sections provide details of detection of these features.

6.2 Point Detection

Point detection in images can be performed using Laplacian masks. These masks have the largest coefficient at the center of the mask while neighbor pixels have an opposite sign. These masks give the highest response to the object that has a similar shape to the mask, such as isolated points. Here, the sum of all coefficients of the mask is equal to zero. This is because the response of the filter must be zero inside a constant intensity area. Examples of Laplacian masks is given in Fig. 6.1 [1].

The SCILAB code for point detection is given here and the result of point detection is shown in Fig. 6.2.

R. M. Thanki, A. M. Kothari, *Digital Image Processing using SCILAB*,
https://doi.org/10.1007/978-3-319-89533-8_6

Fig. 6.1 Laplacian masks for point detection

-1	-1	-1
-1	8	-1
-1	-1	-1

0	-1	0
-1	4	-1
0	-1	0

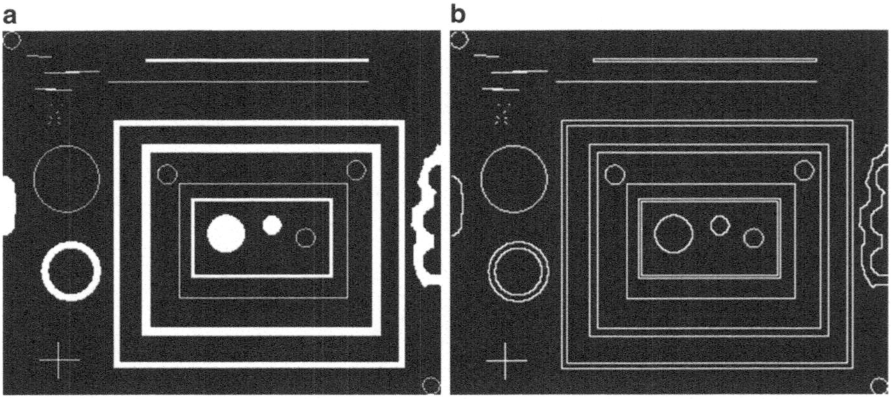

Fig. 6.2 Result of point detection: (**a**) original image; and (**b**) processed image

SCILAB Code for Point Detection

```
z=uigetfile('*.*','Select an image');
a=imread(z);
r=size(a,1);
c=size(a,2);
n=[-1 -1 -1 ; -1 8 -1 ; -1 -1 -1];

for i=2:r-1
  for j=2:c-1
    new(i,j)= a(i-1,j-1)*n(1,1) + a(i-1,j)*n(1,2) + a(i-1,j+1)*n
(1,3) + a(i,j-1)*n(2,1) + a(i,j)*n(2,2) + a(i,j+1)*n(2,3) + a(i+1,
j-1)*n(3,1) + a(i+1,j)*n(3,2) + a(i+1,j+1)*n(3,3);
  end
end
imshow(uint8(new));title('Processed Image');
```

6.3 Line Detection

Line detection in an image is also carried out using Laplacian masks. Four different masks (see Fig. 6.3) are available for line detection in different directions in any image. The lines can be detected in a horizontal, vertical, +45°, and –45° direction [1].

a

-1	-1	-1
2	2	2
-1	-1	-1

b

-1	2	-1
-1	2	-1
-1	2	-1

c

-1	-1	2
-1	2	-1
2	-1	-1

d

2	-1	-1
-1	2	-1
-1	-1	2

Fig. 6.3 Masks for line detection: (**a**) horizontal direction; (**b**) vertical direction; (**c**) +45° direction; and (**d**) –45° direction

The SCILAB code for line detection is given here and the result of line detection is shown in Fig. 6.4.

SCILAB Code for Line Detection

```
z=uigetfile('*.*','Select an image');
a=imread(z);
r=size(a,1);
c=size(a,2);

h=[-1 -1 -1 ; 2 2 2 ; -1 -1 -1];
p=[-1 -1 2 ; -1 2 -1 ; 2 -1 -1];
v=[-1 2 -1 ; -1 2 -1 ; -1 2 -1];
m=[2 -1 -1 ; -1 2 -1 ; -1 -1 2];

for i=2: r-1
  for j=2:c-1
    hor(i,j)=a(i-1,j-1)*h(1) +a(i-1,j)*h(2) +a(i-1,j+1)*h(3) +
a(i,j-1)*h(4) + a(i,j)*h(5) +a(i,j+1)*h(6) +a(i+1,j-1)*h(7) +a
(i+1,j)*h(8) +a(i+1,j+1)*h(9);

    p(i,j)=a(i-1,j-1)*p(1) +a(i-1,j)*p(2) +a(i-1,j+1)*p(3) +a
(i,j-1)*p(4) + a(i,j)*p(5) +a(i,j+1)*p(6) +a(i+1,j-1)*p(7) -a(i
+1,j)*p(8) +a(i+1,j+1)*p(9);

   ver(i,j)=a(i-1,j-1)*v(1) +a(i-1,j)*v(2) +a(i-1,j+1)*v(3) +a
(i,j-1)*v(4) + a(i,j)*v(5) +a(i,j+1)*v(6) +a(i+1,j-1)*v(7) -a(i
+1,j)*v(8) +a(i+1,j+1)*v(9);

    m(i,j)=a(i-1,j-1)*m(1) +a(i-1,j)*m(2) +a(i-1,j+1)*m(3) +a
(i,j-1)*m(4) + a(i,j)*m(5) +a(i,j+1)*m(6) +a(i+1,j-1)*m(7) -a(i
+1,j)*m(8) +a(i+1,j+1)*m(9);

  end
end
subplot(321);imshow(a);title('Original Image');
subplot(323);imshow(hor);title('Horizontal Lines');
subplot(324);imshow(p);title('+45 Degree Lines');
subplot(325);imshow(ver);title('Vertical Lines');
subplot(326);imshow(m);title('-45 Degree Lines');
```

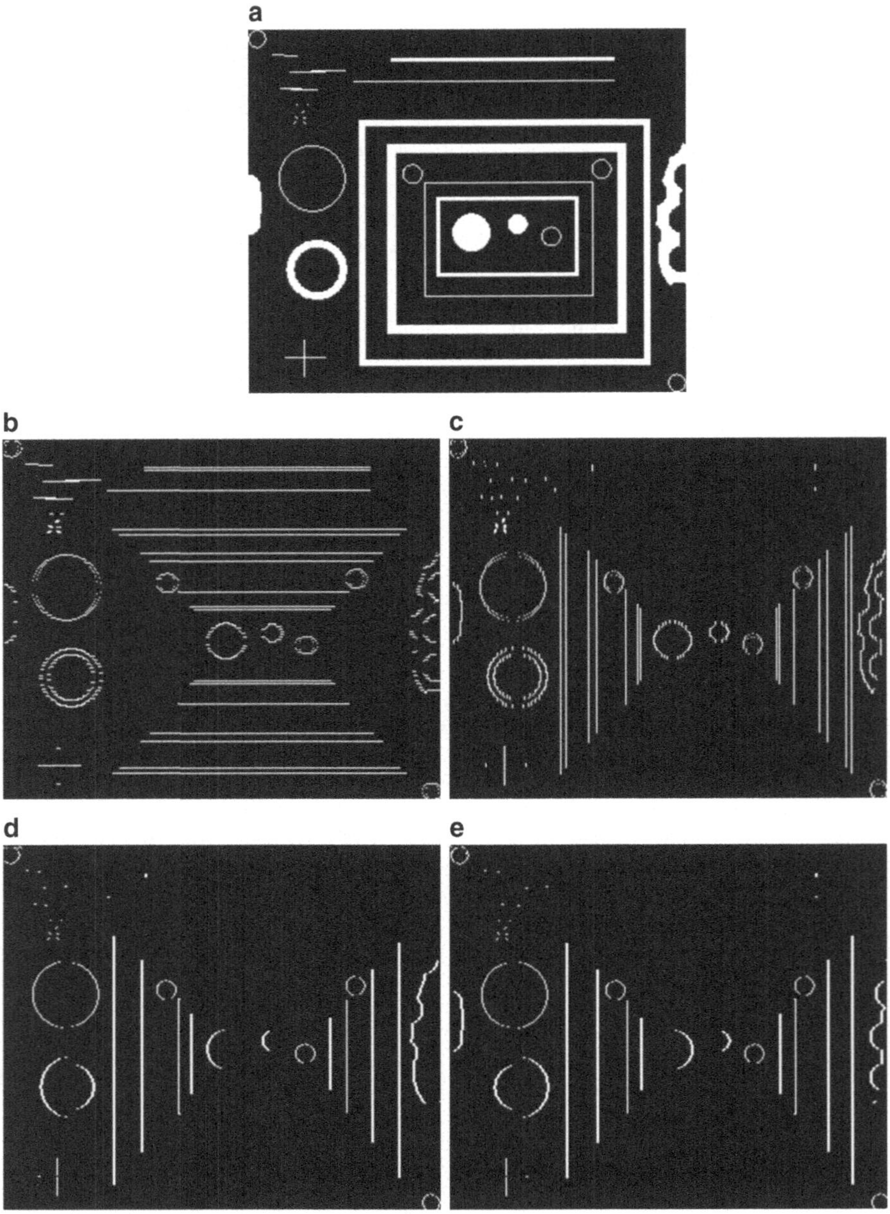

Fig. 6.4 Result of line detection: (**a**) original image; (**b**) horizontal lines; (**c**) vertical lines; (**d**) +45° lines; and (**e**) –45° lines

6.4 Edge Detection

Generally, objects and the background in images have different intensities. Therefore, the edges of objects are the areas in which abrupt intensity changes occur in images. Thus, edge detection is an important process in image segmentation. The edges of an image are detected using first- and second-order derivatives. The gradient of image function $f(x, y)$ is defined as a vector, as follows:

$$\nabla f = \begin{bmatrix} G_x \\ G_y \end{bmatrix} = \begin{bmatrix} \frac{\partial f}{\partial x} \\ \frac{\partial f}{\partial y} \end{bmatrix} \tag{6.1}$$

The magnitude of this vector is:

$$\nabla f = \left[(\partial f/\partial x)^2 + (\partial f/\partial y)^2 \right]^{1/2} \tag{6.2}$$

The fundamental property of the gradient vector is that it points in the direction of the maximum rate of change of the image vector f at coordinates (x, y). The angle at which this maxium rate of change occurs is as follows:

$$\alpha(x, y) = \tan^{-1}\left(\frac{G_y}{G_x}\right) \tag{6.3}$$

In the edge detection algorithm, the assumption is that the edges are the pixels with a high gradient. A fast rate of change of intensity in a direction given by the angle of the gradient vector is observed in edge pixels. The various types of edge detection algorithms are available for extraction of edges in the image. These edge algorithms and their implementation are discussed in Sects. 6.4.1, 6.4.2, 6.4.3, 6.4.4, and 6.4.5.

6.4.1 Sobel Edge Detector

The Sobel edge detector uses the masks [1] shown in Fig. 6.5 to find the first derivatives *Gx* and *Gy*. This edge detector detects edges in horizontal, vertical, or both directions in the image.

Fig. 6.5 Masks for the Sobel edge detector

a

-1	-2	-1
0	0	0
-1	-2	-1

b

-1	0	-1
-2	0	-2
-1	0	-1

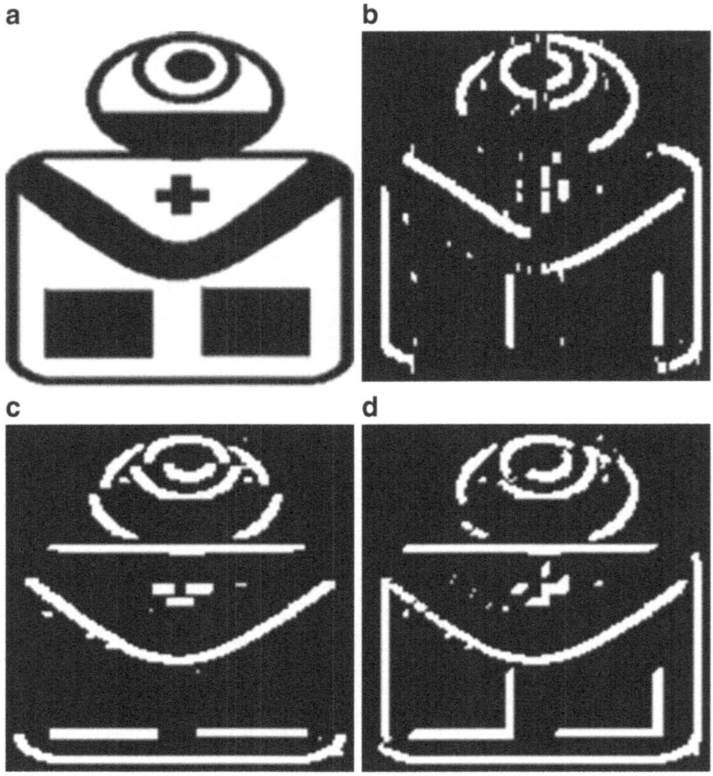

Fig. 6.6 Result from the Sobel edge detector: (**a**) original image; (**b**) edges detected in horizontal direction; (**c**) edges detected in vertical direction; and (**d**) edges detected in both directions

The SCILAB code for the Sobel edge detector is given here and results from the Sobel edge detector are shown in Fig. 6.6.

SCILAB Code for the Sobel Edge Detector

```
a=uigetfile('*.*','Select the Image:-');
a=imread(a);
b=double(a);
d=b;
mh=[-1 -2 -1;0 0 0;1 2 1];
mv=[-1 0 1;-2 0 2;-1 0 1];
[r1,c1]=size(a);
for i=1:r1
  for j=1:c1
    new(i,j)=a(i,j);
    nem(i,j)=a(i,j);
  end
end
new=double(new);
nem=double(nem);
```

```
for i=2:1:r1-1
  for j=2:1:c1-1
    new(i,j)=(mh(1)*d(i-1,j-1))+(mh(2)*d(i-1,j))+(mh(3)*d(i-1,
j+1))......
          +(mh(4)*d(i,j-1))+(mh(5)*d(i,j))+(mh(6)*d(i,j+1))......
        +(mh(7)*d(i+1,j-1))+(mh(8)*d(i+1,j))+(mh(9)*d(i+1,j+1));
  end
end
for i=2:1:r1-1
  for j=2:1:c1-1
    nem(i,j)=(mv(1)*d(i-1,j-1))+(mv(2)*d(i-1,j))+(mv(3)*d(i-1,
j+1))......
         +(mv(4)*d(i,j-1))+ (mv(5)*d(i,j))+(mv(6)*d(i,j+1))......
            +(mv(7)*d(i+1,j-1))+(mv(8)*d(i+1,j))+(mv(9)*d(i+1,j
+1));
  end
end
new2=new+nem;
subplot(221);imshow(uint8(a));title('Original Image');
subplot(222);imshow(uint8(new));title('X-Gradiant Image');
subplot(223);imshow(uint8(nem));title('Y-Gradiant Image');
subplot(224);imshow(uint8(new2));title('Resultant Gradiant
Image');
```

6.4.2 *Prewitt Edge Detector*

The Prewitt edge detector uses the masks [1] shown in Fig. 6.7 to find the first derivatives *Gx* and *Gy*. This detector is easier to implement than the Sobel detector but introduces more noise as a result.

The SCILAB code for the Prewitt edge detector is given here and results from the Prewitt edge detector are given in Fig. 6.8.

SCILAB Code for the Prewitt Edge Detector

```
a=uigetfile('*.*','Select the Image:-');
a=imread(a);
b=double(a);
d=b;
mh=[-1 -1 -1;0 0 0;1 1 1];
mv=[-1 0 1;-1 0 1;-1 0 1];
[r1,c1]=size(a);
for i=1:r1
  for j=1:c1
```

Fig. 6.7 Masks for the Prewitt edge detector

a

-1	-1	-1
0	0	0
-1	-1	-1

b

-1	0	-1
-1	0	-1
-1	0	-1

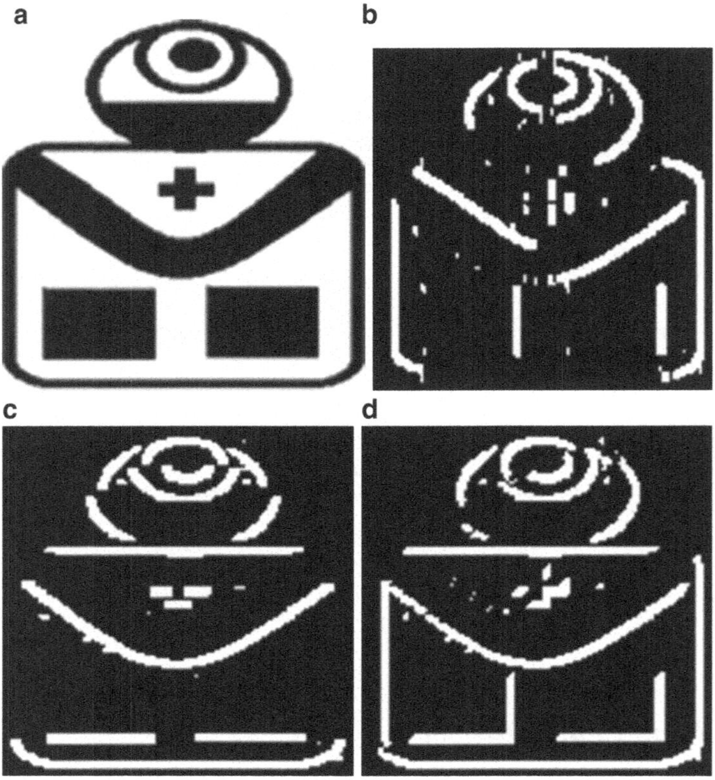

Fig. 6.8 Result from the Prewitt edge detector: (**a**) original image; (**b**) edges detected in horizontal direction; (**c**) edges detected in vertical direction; and (**d**) edges detected in both directions

```
     new(i,j)=a(i,j);
     nem(i,j)=a(i,j);
   end
end
new=double(new);
nem=double(nem);
for i=2:1:r1-1
   for j=2:1:c1-1
     new(i,j)=(mh(1)*d(i-1,j-1))+(mh(2)*d(i-1,j))+(mh(3)*d(i-1,
j+1))......
        +(mh(4)*d(i,j-1))+(mh(5)*d(i,j))+(mh(6)*d(i,j+1))......
       +(mh(7)*d(i+1,j-1))+(mh(8)*d(i+1,j))+(mh(9)*d(i+1,j+1));
   end
end
for i=2:1:r1-1
   for j=2:1:c1-1
     nem(i,j)=(mv(1)*d(i-1,j-1))+(mv(2)*d(i-1,j))+(mv(3)*d(i-1,
j+1))......
       +(mv(4)*d(i,j-1))+ (mv(5)*d(i,j))+(mv(6)*d(i,j+1))......
       +(mv(7)*d(i+1,j-1))+(mv(8)*d(i+1,j))+(mv(9)*d(i+1,j+1));
```

Fig. 6.9 Masks for the Roberts edge detector

a

-1	0
0	1

b

-1	0
0	1

```
  end
end
new2=new+nem;
subplot(221);imshow(uint8(a));title('Original Image');
subplot(222);imshow(uint8(new));title('X-Gradiant Image');
subplot(223);imshow(uint8(nem));title('Y-Gradiant Image');
subplot(224);imshow(uint8(new2));title('Resultant Gradiant
Image');
```

6.4.3 Roberts Edge Detector

The Roberts edge detector is the simplest edge detector algorithm [1–3], which was proposed and implemented by Lawrence Roberts in 1963 [3]. This edge detector uses the masks shown in Fig. 6.9 to find the first derivatives Gx and Gy.

The SCILAB code for the Roberts edge detector is given here and the results of the Roberts edge detector are shown in Fig. 6.10.

SCILAB Code for the Roberts Edge Detector

```
a=uigetfile('*.*','Select the Image:-');
a=imread(a);
b=double(a);
d=b;
mh=[1 0;0 -1];
mv=[0 -1;1 0];
[r1,c1]=size(a);
for i=1:r1
  for j=1:c1
    new(i,j)=a(i,j);
    nem(i,j)=a(i,j);
  end
end
new=double(new);
nem=double(nem);
for i=2:1:r1-1
  for j=2:1:c1-1
    new(i,j)=(mh(1)*d(i,j))+(mh(2)*d(i,j+1))......
        +(mh(3)*d(i+1,j))+(mh(4)*d(i+1,j+1));
  end
end
for i=2:1:r1-1
  for j=2:1:c1-1
    nem(i,j)=(mv(1)*d(i,j))+(mv(2)*d(i,j+1))......
        +(mv(3)*d(i+1,j))+(mv(4)*d(i+1,j+1));
```

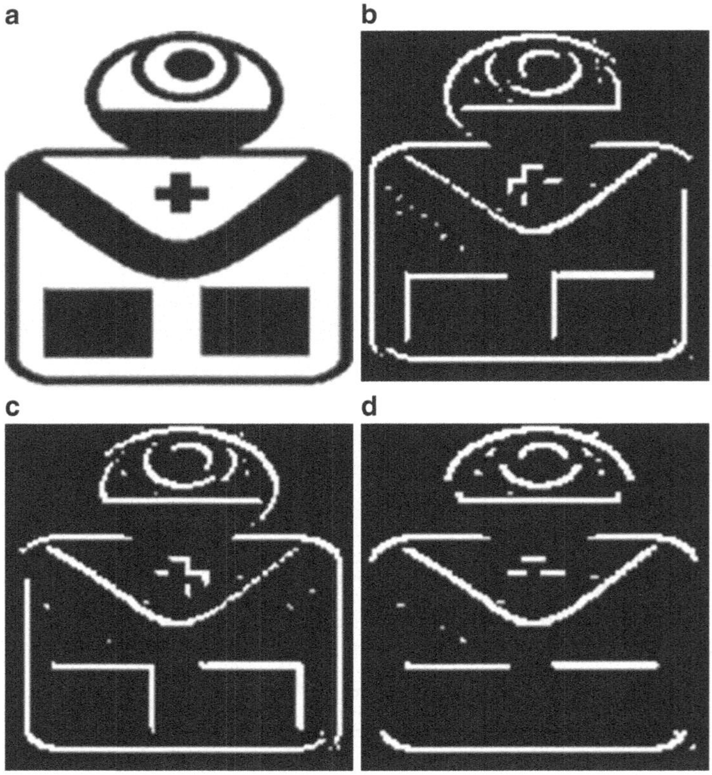

Fig. 6.10 Results of the Roberts edge detector: (**a**) original image; (**b**) edges detected in horizontal direction; (**c**) edges detected in vertical direction; and (**d**) edges detected in both directions

```
  end
end
new2=new+nem;

subplot(221);imshow(uint8(a));title('Original Image');
subplot(222);imshow(uint8(new));title('X-Gradiant Image');
subplot(223);imshow(uint8(nem));title('Y-Gradiant Image');
subplot(224);imshow(uint8(new2));title('Resultant Gradiant
Image');
```

6.4.4 Laplacian of a Gaussian (LOG) Edge Detector

Consider the following Gaussian function:

$$h(a) = -e^{-\frac{a^2}{2\sigma^2}} \tag{6.4}$$

where $a^2 = x^2 + y^2$ and σ^2 is the standard deviation.

This is smoothing function which introduces blurring in the image. The effect of blurring depends on strength of standard deviation. The Laplacian of Eq. 6.4 is given as follows:

$$\nabla^2 h(a) = -\left[\frac{a^2 - \sigma^2}{\sigma^4}\right] e^{-\frac{r^2}{2\sigma^2}} \tag{6.5}$$

The function shown in Eq. 6.5 is referred to as the Laplacian of Gaussian (LOG). This function has two effects on an image: it smoothes the image that is corrupted by noise and it computes the Laplacian, which gives a double-edged image. The SCILAB code for the LOG edge detector is given here and the results of the LOG edge detector are shown in Fig. 6.11.

SCILAB Code for the Laplacian of a Gaussian Edge Detector

```
a=uigetfile('*.*','Select the Image:-');
a=imread(a);
b=double(a);
d=b;
mh=[0,-1,0;1,-4,1;0,1,0];
mv=[0,0,-1,0,0;0,-1,-2,-1,0;-1,-2,16,-2,-1;0,-1,-2,-1,0;0.0,-
1,0,0];
[r1,c1]=size(a);
for i=1:r1
  for j=1:c1
    LAP(i,j)=a(i,j);
    LoG(i,j)=a(i,j);
  end
end
new=double(LAP);
nem=double(LoG);
for i=3:1:r1-2
  for j=3:1:c1-2
    new(i,j)=(mh(1)*d(i-1,j-1))+(mh(2)*d(i-1,j))+(mh(3)*d(i-1,j
+1))......
       +(mh(4)*d(i,j-1))+(mh(5)*d(i,j))+(mh(6)*d(i,j+1))......
       +(mh(7)*d(i+1,j-1))+(mh(8)*d(i+1,j))+(mh(9)*d(i+1,j+1));
  end
end
for i=3:1:r1-2
  for j=3:1:c1-2
    nem(i,j)=(mv(1)*d(i-2,j-2))+(mv(2)*d(i-2,j-1))+(mv(3)*d
(i-2,j))+(mv(4)*d(i-2,j+1))+(mv(5)*d(i-2,j+2))......
```

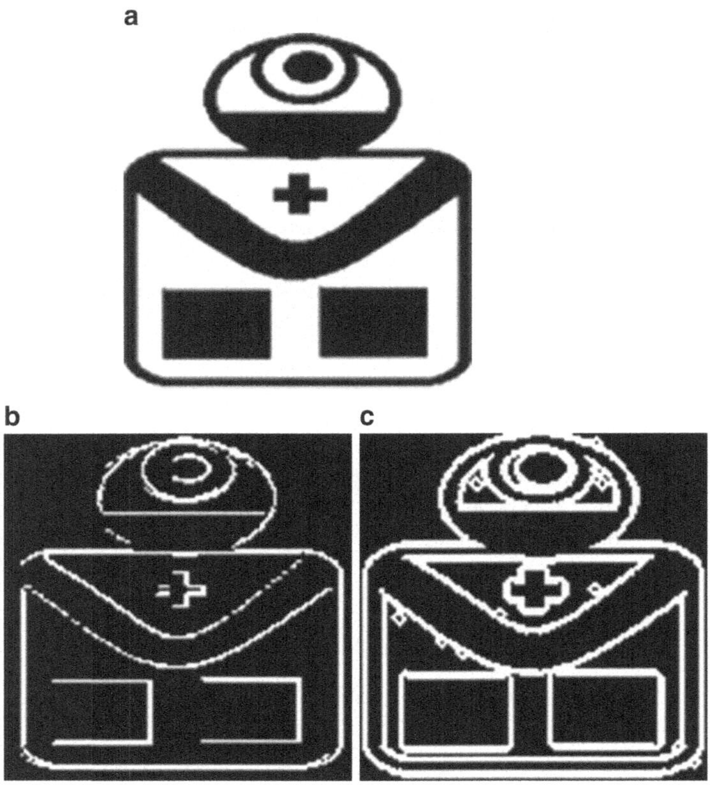

Fig. 6.11 Results of the Laplacian of a Gaussian (LOG) edge detector: (**a**) original image; (**b**) Laplacian image; and (**c**) Laplacian of a Gaussian image

```
        +(mv(6)*d(i-1,j-2))+(mv(7)*d(i-1,j-1))+(mv(8)*d(i-1,j))
+(mv(9)*d(i-1,j+1))+(mv(10)*d(i-1,j+2)).....
         +(mv(11)*d(i,j-2))+(mv(12)*d(i,j-1))+(mv(13)*d(i,j))
+(mv(14)*d(i,j+1))+(mv(15)*d(i,j+2)).....
        +(mv(16)*d(i+1,j-2))+(mv(17)*d(i+1,j-1))+(mv(18)*d(i+1,
j))+(mv(19)*d(i+1,j+1))+(mv(20)*d(i+1,j+2)).....
        +(mv(21)*d(i+2,j-2))+(mv(22)*d(i+2,j-1))+(mv(23)*d(i+2,
j))+(mv(24)*d(i+2,j+1))+(mv(25)*d(i+2,j+2));
 end
end
subplot(131);imshow(uint8(a));title('Original Image');
subplot(132);imshow(uint8(new));title('Laplacian Image');
subplot(133);imshow(uint8(nem));title('Laplacian of Gaussian
Image');
```

6.4.5 Canny Edge Detector

The Canny edge detector algorithm was proposed and implemented by John F. Canny in 1986 [1, 2, 4]. It detects edges of images using a Gaussian filter predefined with a standard deviation value. The steps for the Canny edge detection algorithm are as follows [4, 5]:

1. Read the image *I*.
2. Convolve a one-dimensional (1D) Gaussian mask with image *I*.
3. Create a 1D mask for the first derivative of the Gaussian in the *x* direction and *y* direction.
4. Convolve image *I* with *G* along the row to obtain *Ix* and down the columns to obtain *Iy*.
5. Convolve *Ix* with *Gx* to get *Ix′* and *Iy* with *Gy* to get *Iy′*.
6. Find the magnitude of the result at each pixel (*x*, *y*) using the following equation:

$$M(x,y) = \sqrt{I'_x(x,y)^2 + I''_y(x,y)^2} \tag{6.6}$$

There is an inbuilt function 'Canny' edge algorithm in SCILAB and the syntax for this algorithm is given as follows:

$$\mathbf{E = edge(im, 'canny', thresh, sigma)};$$

where im is an input image, thresh is a threshold value and cannot be negative (default value is 0.2), and sigma is an aperture parameter for the Sobel operator (default value is 3). The simple SCILAB code for edge detection using canny edge detector is given here and the result is shown in Fig. 6.12.

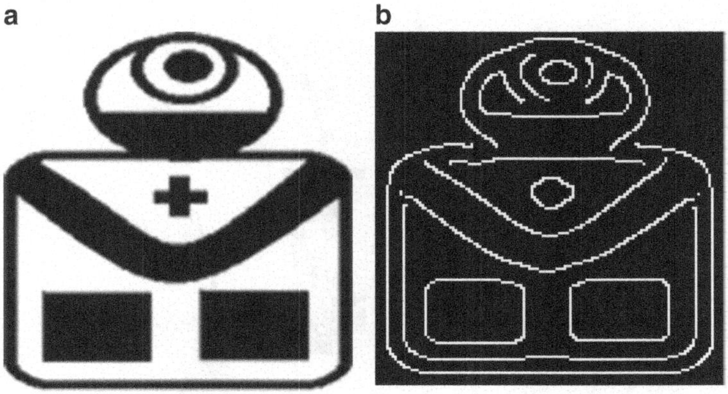

Fig. 6.12 Result from the Canny edge detector: (**a**) original image; and (**b**) processed image

SCILAB Code for Canny Edge Detector

```
a=uigetfile('*.*','Select the Image:-');
a=imread(a);
b=double(a);
thresh=0.4;
sigma=3;
E=edge(a, 'canny', thresh, sigma);
imshow(E);
```

6.5 Thresholding

For simplicity of implementation of the image processing applications, the thresholding process is the first step for image enhancement. Suppose image $f(x, y)$ has white objects on a dark background, in such a way that object pixels and background pixels have intensity levels grouped into two types. This process of extracting objects from the background is to select a threshold T that separates these two types. Then for any point (x, y) for f where $f(x, y) \geq T$ is called the object point. Otherwise, the point is called the background point. The threshold image $g(x, y)$ can be given as follows:

$$\begin{aligned} g(x,y) &= 1, f(x,y) \geq T \\ g(x,y) &= 0, f(x,y) < T \end{aligned} \tag{6.7}$$

If the value of threshold T is constant it is called global thresholding. If the value of threshold T is variable it is called local thresholding. The simple SCILAB code for global thresholding is given here and the result is shown in Fig. 6.13.

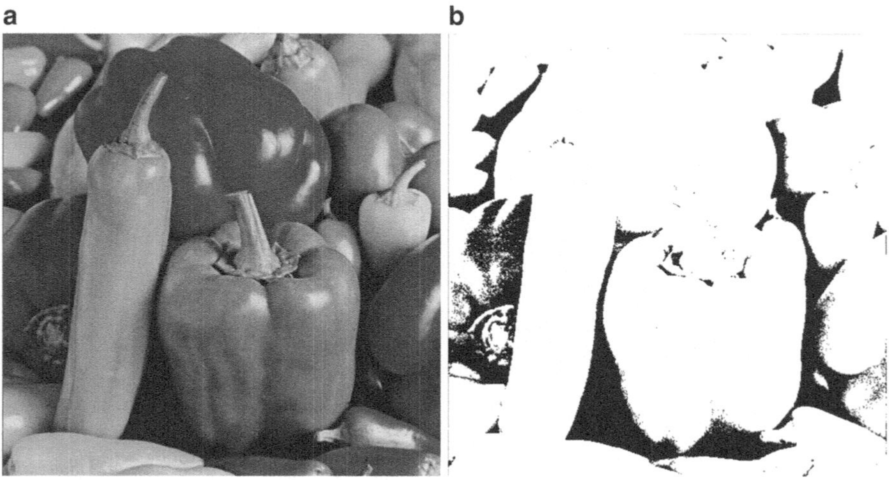

Fig. 6.13 Result of global thresholding: (**a**) original image; and (**b**) processed image

SCILAB Code for Global Thresholding

```
a=uigetfile('*.*','Select the Image');
a=imread(a);

th=58;

[r,c]=size(a);

for i=1:r
  for j=1:c

    if (a(i,j)<=th)
      x(i,j)=0;
    else
      x(i,j)=255;
    end
  end
end
x=uint8(x);
diff=imabsdiff(a,x);
new=x;
imshow(new);title('Thresholding Image');
```

Bibliography

1. Gonzalez, R., & Woods, R. (2008). *Digital image processing*. Upper Saddle River: Pearson Education India.
2. Gonzalez, R., Woods, R., & Eddins, L. (2009). *Digital image processing using MATLAB*. New Delhi: TATA McGraw-Hill Education.
3. Roberts, L. G. (1965) Machine perception of three-dimensional solids. Massachusetts Institute of Technology, USA. http://www.packet.cc/files/mach-per-3D-solids.html
4. Canny, J. (1986). A computational approach to edge detection. *IEEE Transaction on Pattern Analysis and Machine Intelligence, 8*(6), 679–698.
5. Trivedi, D., Kothari, M., Shah, S., & Shingala, N. (2014). Dental image matching by Canny algorithm for human identification. *International Journal of Advanced Computer Research, 4* (17), 985–990.

Chapter 7
Color Image Processing

7.1 Color Image Representation in SCILAB

A color image is a group of color pixels with size of $M \times N \times 3$, where each color pixel is a triplet combination of the red, green, and blue color of a color image. The color image can be viewed as three gray-scale images that, when fed into the red, green, and blue channels of a color monitor produce a color image on the screen. The three images obtained a color image are called the red, green, and blue component images. The color image is defined by a data class of uint8 or uint16. If the color image is in the double class its range of values is [0, 1]. The number of bits required to represent the pixel values of color component images determines the bit depth of the color image [1, 2]. If each color component image is an 8-bit image the color image is has a 24-bit depth. Generally, the bit depth of each component image is the same. Therefore, the bit depth of a color image is given as $(2^b)^3$, where b is the number of bits. An 8-bit color image has 16,777,216 colors [1, 2].

The syntax for displaying color component images of color images in SCILAB is given here and images are shown in Fig. 7.1.

```
a=uigetfile('*.*','Select the Image:-');
a=imread(a);
Red_Component=a(:,:,1);
Green_Component=a(:,:,2);
Blue_Component=a(:,:,3);

subplot(131);imshow(Red_Component);title('Red Component
Image');
subplot(132);imshow(Green_Component);title('Green Component
Image');
subplot(133);imshow(Blue_Component);title('Blue Component
Image');
```


R. M. Thanki, A. M. Kothari, *Digital Image Processing using SCILAB*,
https://doi.org/10.1007/978-3-319-89533-8_7

Fig. 7.1 Displaying color images: (**a**) original image; (**b**) red component image; (**c**) green component image; and (**d**) blue component image

7.2 Conversion of an RGB (Red, Green, Blue) Image into Other Spaces

Any color image can be represented by various types of color spaces. These color spaces are the NTSC (National Television System Committee) color space, YCbCr (luminance, chrominance) color space, HSV (hue, saturation, value) color space, and CMY (cyan, magenta, yellow) color space [1–2]. RGB (red, green, blue) color images can be converted into any of these color spaces using SCILAB. The inbuilt functions are available for conversion of a color image into any spaces and vice versa. Details on the conversion of a color image into different color spaces are given in Sects. 7.2.1, 7.2.2, 7.2.3, and 7.2.4.

7.2.1 NTSC Color Space

The NTSC color space system was introduced in the USA for video transmission in the television communication system [3, 4]. In this color space, a color image has three components: luminance (Y), hue (I), and saturation (Q). This color space is also known as the YIQ color space. Any RGB color image can be converted in YIQ color space using Eq. 7.1:

$$\begin{aligned} Y &= 0.299 \cdot R + 0.587 \cdot G + 0.114 \cdot B \\ I &= 0.596 \cdot R - 0.274 \cdot G - 0.322 \cdot B \\ Q &= 0.211 \cdot R - 0.523 \cdot G + 0.312 \cdot B \end{aligned} \tag{7.1}$$

Any color image in the YIQ color space can be converted into the RGB color space using Eq. 7.2:

$$\begin{aligned} R &= Y + 0.956 \cdot I + 0.621 \cdot Q \\ G &= Y - 0.272 \cdot I - 0.647 \cdot Q \\ B &= Y - 1.106 \cdot I + 1.703 \cdot Q \end{aligned} \tag{7.2}$$

The SCILAB function syntax for the RGB color space to the NTSC color space and NTSC to RGB are as follows:

```
YIQ = rgb2ntsc(RGB);
RGB = ntsc2rgb(YIQ);
```

The simple SCILAB code for an RGB color image converted into the NTSC color space is given here and the result is shown in Fig. 7.2.

```
I=uigetfile('*.*','Select the Image');
image=imread(I);
YIQ = rgb2ntsc(image);
imwrite(YIQ,'NTSC.tif');
subplot(121);imshow(YIQ);title('Image in NTSC Color Space');
```

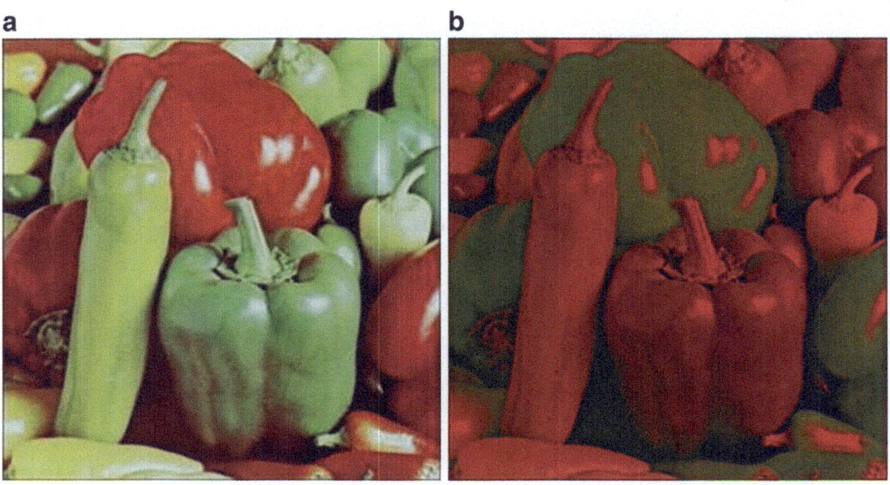

Fig. 7.2 (**a**) Original image. (**b**) Image in the NTSC color space

7.2.2 YCbCr Color Space

The YCbCr color space is used mainly in applications related to video. In this space, the luminance component of an image is represented by the component *Y*, and color information is represented by the two components Cb and Cr. Cb is the difference value between the blue channel value and a reference value, and Cr is the difference value between the red channel value and a reference value [1–2]. Any RGB color image can be converted into the YCbCr color space using Eq. 7.3:

$$\begin{aligned} Y &= 0.299 \cdot R + 0.587 \cdot G + 0.114 \cdot B \\ \mathrm{Cb} &= 0.564 \times (B - Y) + 0.5 \\ \mathrm{Cr} &= 0.713 \times (R - Y) + 0.5 \end{aligned} \tag{7.3}$$

In addition, any color image in the YCbCr color space can be converted into the RGB color space using Eq. 7.4:

$$\begin{aligned} R &= Y + 1.403 \times (\mathrm{Cr} - 0.5) \\ G &= Y - 0.344 \times (\mathrm{Cr} - 0.5) - 0.714 \times (\mathrm{Cb} - 0.5) \\ B &= Y + 1.773 \times (\mathrm{Cb} - 0.5) \end{aligned} \tag{7.4}$$

The SCILAB function syntax for an RGB color space converted to the YCbCr color space and YCbCr to RGB are as follows:

```
YCC = rgb2ycbcr(RGB);
RGB = ycbcr2rgb(YCC);
```

The simple SCILAB code for a RGB color image converted into the YCbCr color space is given here and the result is shown in Fig. 7.3.

```
I=uigetfile('*.*','Select the Image');
image=imread(I);
YCC = rgb2ycbcr(image);
imwrite(YCC,'YCbCr.tif');
subplot(121);imshow(YCC);title('Image in YCbCr Color Space');
```

7.2.3 HSV (Hue, Saturation, Value) Color Space

HSV is a color space that is used by users to choose a specific color from a color palette. This color space is closer to the RGB color space in which a user describes a color based on experience. Any RGB color image can be converted into the HSV color space using Eq. 7.5:

Fig. 7.3 (**a**) Original image. (**b**) Image in the YCbCr color space

$$\begin{aligned}&V = \max(R, G, B)\\&S = (V - \min(R, G, B))/V, if\ \ V <> 0\\&H = (G - B)/6/S, V = R;\\&H = 1/2 + (B - R)/6/S, V = G;\\&H = 2/3 + (R - G)/6/S, V = B.\end{aligned} \tag{7.5}$$

The SCILAB function syntax for an RGB color space converted into the HSV color space and HSV to RGB are as follows:

```
HSV = rgb2hsv(RGB);
RGB = hsv2rgb(HSV);
```

The simple SCILAB code for an RGB color image converted into the HSV color space is given here and the result is shown in Fig. 7.4.

```
I=uigetfile('*.*','Select the Image');
image=imread(I);
HSV = rgb2hsv(image);
imwrite(HSV,'HSV.tif');
subplot(121);imshow(HSV);title('Image in HSV Color Space');
```

7.2.4 *CMY (Cyan, Magenta, Yellow) Color Space*

The CMY color space represents an image using secondary colors of lights such as cyan, magenta, and yellow. The SCILAB function syntax for converting the RGB color space to the CMY color space is as follows:

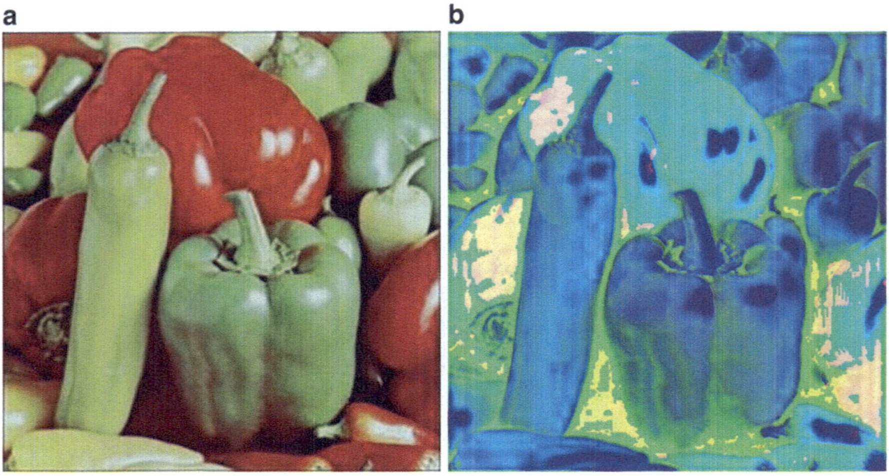

Fig. 7.4 (a) Original image. (b) Image in the HSV (hue, saturation, value) color space

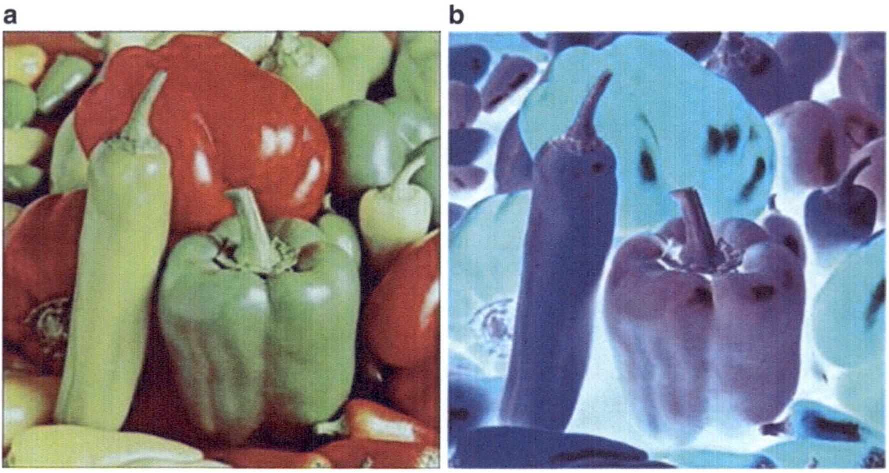

Fig. 7.5 (a) Original image. (b) Image in the CMY (cyan, magenta, yellow) color space

```
CMY = imcomplement(RGB);
```

The simple SCILAB code for an RGB color image converted into the CMY color space is given here and the result is shown in Fig. 7.5.

```
I=uigetfile('*.*','Select the Image');
image=imread(I);
CMY = imcomplement(image);
imwrite(CMY,'CMY.tif');
subplot(121);imshow(CMY);title('Image in CMY Color Space');
```

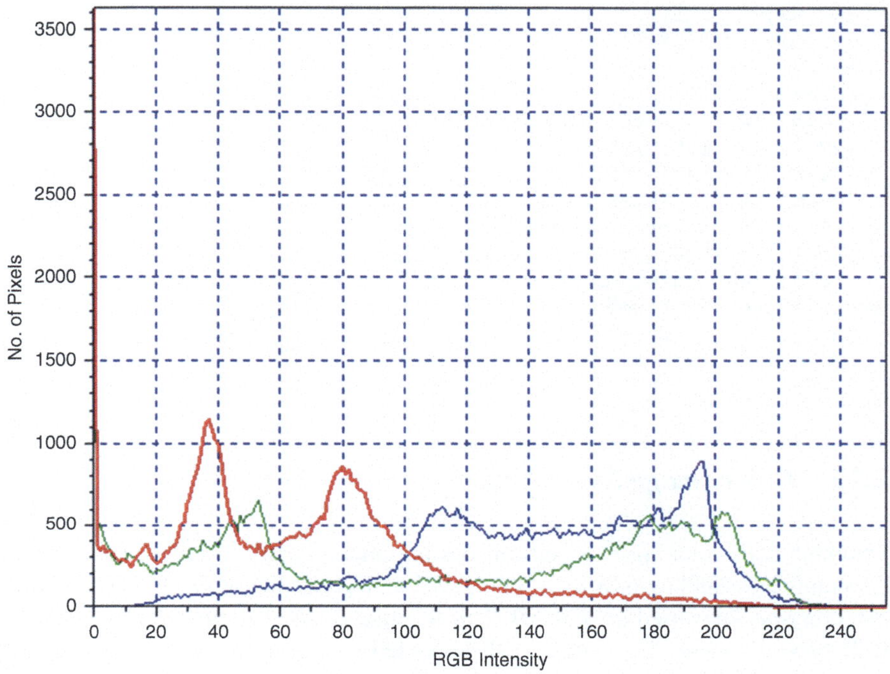

Fig. 7.6 Histogram of color image

7.3 Basic Operations for a Color Image

In this section, we discuss some basic SCILAB codes for displaying histograms of color images and various filter applications for color images are given.

7.3.1 Histogram of a Color Image

In a gray-scale image, the histogram gives the number of pixels in a particular gray level, but in an RGB color image, the histogram gives the number of pixels in a particular color value. In the RGB color image, any pixel has a combination of three colors (red, green, and blue). Thus, we have to plot an individual histogram of each component of the RGB color image. The simple SCILAB code for a histogram of the color image is given here and the result is shown in Fig. 7.6.

SCILAB Code for a Histogram of a Color Image

```
I = uigetfile("*.*","Select the Image");
image = imread(I);
//Split into RGB Components
```

```
R = image(:,:,1);
G = image(:,:,2);
B = image(:,:,3);
nBins = 256;
//Get histValues for each channel
[yR,x] = imhist(R,nBins);
[yG,x] = imhist(G,nBins);
[yB,x] = imhist(B,nBins);
//Plot them together in one plot
plot(x,yR,x,yG,x,yB,"Linewidth",2);
xlabel("RGB Intensity");
ylabel("No. of Pixels");
set(gca(),"grid",[1,1]);
```

7.3.2 Color Image Smoothing

Smoothing of a color image is performed using smoothing filters such as the average filter and Gaussian low-pass filter. In Fig. 7.7 we have demonstrated how the inbuilt function *fspecial* is used in SCILAB. This function is used to define any image filter with a specific value. The simple SCILAB code for color image smoothing is given here and the results are shown in Fig. 7.7.

SCILAB Code for Color Image Smoothing

```
I=uigetfile('*.*','Select the Image');
image = imread(I);
I=imnoise(image,'salt & pepper',0.2);

F = fspecial('average',[3, 3]);
pic(:,:,1) = imfilter(I(:,:,1),F);
pic(:,:,2) = imfilter(I(:,:,2),F);
pic(:,:,3) = imfilter(I(:,:,3),F);
Average_Image = uint8(pic);

F1 = fspecial('gaussian',[3, 3],0.5);
pic1(:,:,1) = imfilter(I(:,:,1),F1);
pic1(:,:,2) = imfilter(I(:,:,2),F1);
pic1(:,:,3) = imfilter(I(:,:,3),F1);
GLPF_Image = uint8(pic1);

imwrite(Average_Image,'Average.tif');
imwrite(GLPF_Image,'GLPF.tif');

subplot(131);imshow(I);title('Noisy Image');
subplot(132);imshow(Average_Image);title('Average Filtered
Image');
subplot(133);imshow(GLPF_Image);title('GLPF Filtered Image');
```

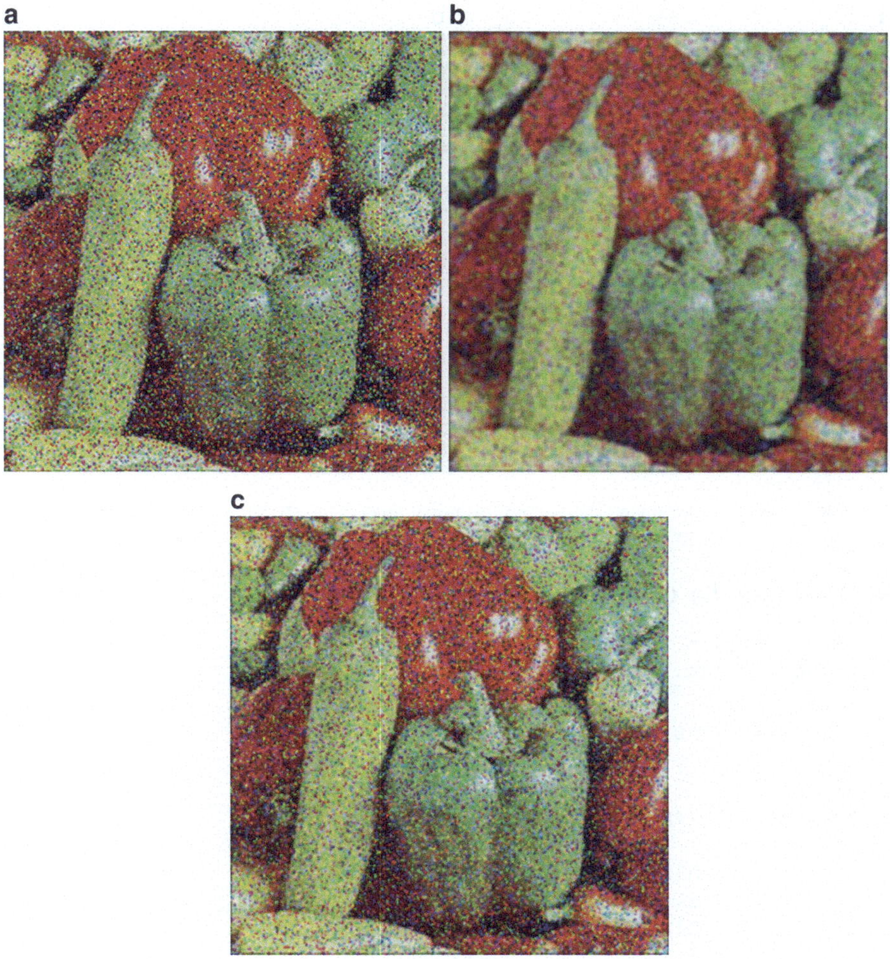

Fig. 7.7 Results of color image smoothing: (**a**) noisy image; (**b**) average filtered image; and (**c**) Gaussian low-pass filter (GLPF) filtered image

7.3.3 Color Image Sharping

Sharpening of a color image using a linear spatial filter is the same as sharpening a gray-scale image. In Fig. 7.8 we have performed sharpening of each component of the color image.

The simple SCILAB code for the color image sharpening is given here and the result is shown in Fig. 7.8.

Fig. 7.8 Result of color image sharpening: (**a**) noisy image; and (**b**) sharpening filtered image

SCILAB Code for Color Image Sharpening

```
I=uigetfile('*.*','Select the Image');
image = imread(I);

I=imnoise(image,'salt & pepper',0.2);

alpha=0.2;
F=fspecial('laplacian',alpha);

pic(:,:,1) = filter2(F,I(:,:,1));
pic(:,:,2) = filter2(F,I(:,:,2));
pic(:,:,3) = filter2(F,I(:,:,3));

Sharp_Image = uint8(pic);

imwrite(uint8(I),'Noisy.tif');
imwrite(uint8(pic),'Sharp.tif');

subplot(121);imshow(I);title('Original Image');
subplot(122);imshow(Sharp_Image);title('Sharp Filtered
Image');
```

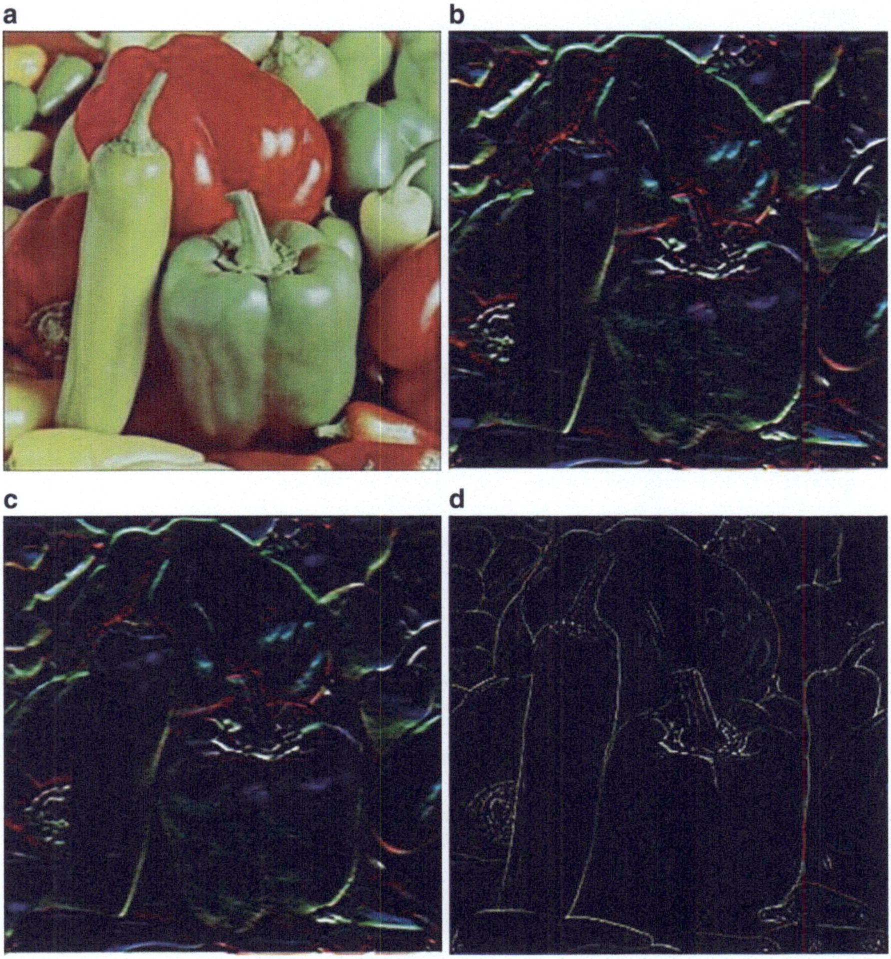

Fig. 7.9 Result of color edge detection: (**a**) original image; (**b**) edges detected by Sobel; (**c**) edges detected by Prewitt; and (**d**) edges detected by Laplacian of a Gaussian (LOG)

7.3.4 Color Edge Detection

In Fig. 7.9 we have demonstrated edge detection of a color image. The inbuilt SCILAB function for the edge detection algorithms is used to extract the edges of a color image. The SCILAB code for color edge detection is given here and the result is shown in Fig. 7.9.

SCILAB Code for Color Edge Detection

```
I=uigetfile('*.*','Select the Image');
image = imread(I);
```

```
hsize = [3, 3];
sigma = 0.5;

F1 = fspecial('sobel');
F2 = fspecial('prewitt');
F3 = fspecial('log',hsize,sigma);

Image1 = imfilter(image, F1);
Image2 = imfilter(image, F2);
Image3 = imfilter(image, F3);

imwrite(uint8(Image1),'Sobel Edges.tif');
imwrite(uint8(Image2),'Prewitt Edges.tif');
imwrite(uint8(Image3),'LOG Edges.tif');

subplot(131);imshow(Image1);title('Sobel Edges');
subplot(132);imshow(Image2);title('Prewitt Edges');
subplot(133);imshow(Image3);title('LOG Edges');
```

Bibliography

1. Gonzalez, R., & Woods, R. (2008). *Digital image processing*. Delhi: Pearson Education India.
2. Gonzalez, R., Woods, R., & Eddins, L. (2009). *Digital image processing using MATLAB*. Delhi: TATA McGraw-Hill Education.
3. Poynton, C. (2012). *Digital video and HD, Chapter 14* (2nd ed, p. 134). Burlington: Morgan Kaufmann.
4. TV Transmission Standards. (2014). U. S. Government Printing Office, p. 210.

Chapter 8
Applications of Digital Image Processing

8.1 Introduction

In the real world, digital image processing is a very effective tool for solving various problems. Image processing is widely used in industry to find defects in products, for packing products, and for identification of objects. It is also used for identification of people and other problems such as traffic, and so on, which we face in daily life. Thus, image processing is a potential area that could help solve many problems arising in today's world. In this chapter, a few applications of image processing such as copyright protection, image compression, fog removal, and template matching are given.

8.2 Copyright Protection

Digital watermarking is widely used for copyright protection of multimedia data such as digital images, videos, and audio signals against unauthenticated distribution over the internet [1–4]. There are various types of watermarking techniques discussed in the literature. Watermarking is a process of embedding the owner's or user's secret information to cover data. The watermarking process is divided into two: the embedding process and extraction process. In the embedding process, the inputs are cover data and watermark data, whereas the output is the watermarked data. In the extraction process, the input is the watermarked data and the output is the extracted watermarked data.

The watermarking technique is designed using various types of multimedia as cover data and, based on multimedia data, watermarking can be classified as image watermarking, video watermarking, audio watermarking, and document watermarking. The watermarking technique can be implemented using various approaches such as least significant bit (LSB) substitution in the spatial domain,

R. M. Thanki, A. M. Kothari, *Digital Image Processing using SCILAB*,
https://doi.org/10.1007/978-3-319-89533-8_8

Fig. 8.1 Results of watermarking: (**a**) original cover image; (**b**) watermark image; (**c**) watermarked image; and (**d**) recovered watermark image

and using various image transforms such as discrete cosine transform (DCT), discrete wavelet transform (DWT), singular value decomposition (SVD), and so on, in the transform domain. The simple SCILAB codes for LSB substitution-based watermarking [5] are given here and the results are shown in Fig. 8.1.

SCILAB Code for the Embedding Process

```
cover_object = uigetfile('*.*','Select the image File');
cover_ooject = imread(cover_object);

message = uigetfile('*.*','Select the image File');
message = imread(message);

message = double(message);
message = round(message ./256);

message = message;
```

```
// determine size of cover object
Mc = size(double(cover_object),1);//Height
Nc = size(double(cover_object),2);//Width

// determine size of message object
Mm = size(message,1);//Height
Nm = size(message,2);//Width

// title the message object out to cover object size to generate
watermark
for ii = 1:Mc
 for jj = 1:Nc
  watermark(ii,jj) = message(pmodulo(ii,Mm)+1,pmodulo(jj,Nm)+1);
 end;
end;

// now we set the lsb of cover_object(ii,jj) to the value of
watermark(ii,jj)
watermarked_image = cover_object;

for ii = 1:Mc
 for jj = 1:Nc

  watermarked_image(ii,jj) = bitset(watermarked_image(ii,jj),1,
watermark(ii,jj));

 end;
end;

watermarked_image_int = uint8(watermarked_image);
// write the watermarked image out to a file
imwrite(watermarked_image_int,'lsb_watermarked.bmp');

imshow(watermarked_image_int);title('Watermarked Image');
```

SCILAB Code for the Extraction Process

```
watermarked_image = imread('lsb_watermarked.bmp');

Mw = size(mtlb_double(watermarked_image),1);//Height
Nw = size(mtlb_double(watermarked_image),2);//Width

for ii = 1:Mw
 for jj = 1:Nw

   watermark(ii,jj) = bitget(watermarked_image(ii,jj),1);

 end;
end;
```

```
watermark = 256*double(mtlb_double(watermark));
imwrite(watermark,'Extracted_watermark.tif');
imshow(watermark);title('Recovered Watermark');
```

8.3 Image Compression

When any image is transmitted through a communication channel, compression of the image is required due to the band limitation of the channel. Therefore, various compression standards are used for compression of images [6–9]. The two most popular image compression standards are joint photographic expert group (JPEG) [8] and JPEG 2000 [9]. The JPEG standard is designed and implemented using DCT and the JPEG 2000 standard is designed and implemented using DWT. In these standards, unwanted frequency coefficients of images are abandoned and the image is reconstructed using the remaining coefficients with a predefined threshold. The low rank approximation property of SVD is also used for image compression. Recently, hybrid compression methods have been designed by researchers using a combination of two or more than two image transforms. These methods are DCT + DWT and DCT + DWT + SVD. The compressed images using these methods are given in Fig. 8.2.

8.4 Fog Removal

In winter, a lot of fog is present in the environment during early morning or throughout the day. Many accidents occur due to fog on highways or streets. An image-processing algorithm can be used for fog removal in captured images and to improve visibility of objects or information in the image. This algorithm can be designed and implemented using a combination of simple image enhancement theories such as histogram equalization, contrast stretching, thresholding, and gamma correction. Figure 8.3 shows a simple example of fog removal from the captured image.

8.5 Template Matching

Template matching is used to identify an object in an image [10]. This application is widely used to find the year of printing in currency, for object detection on a production line, and for hand-written letter recognition. The algorithm for this application can be designed and implemented using the simple correlation function

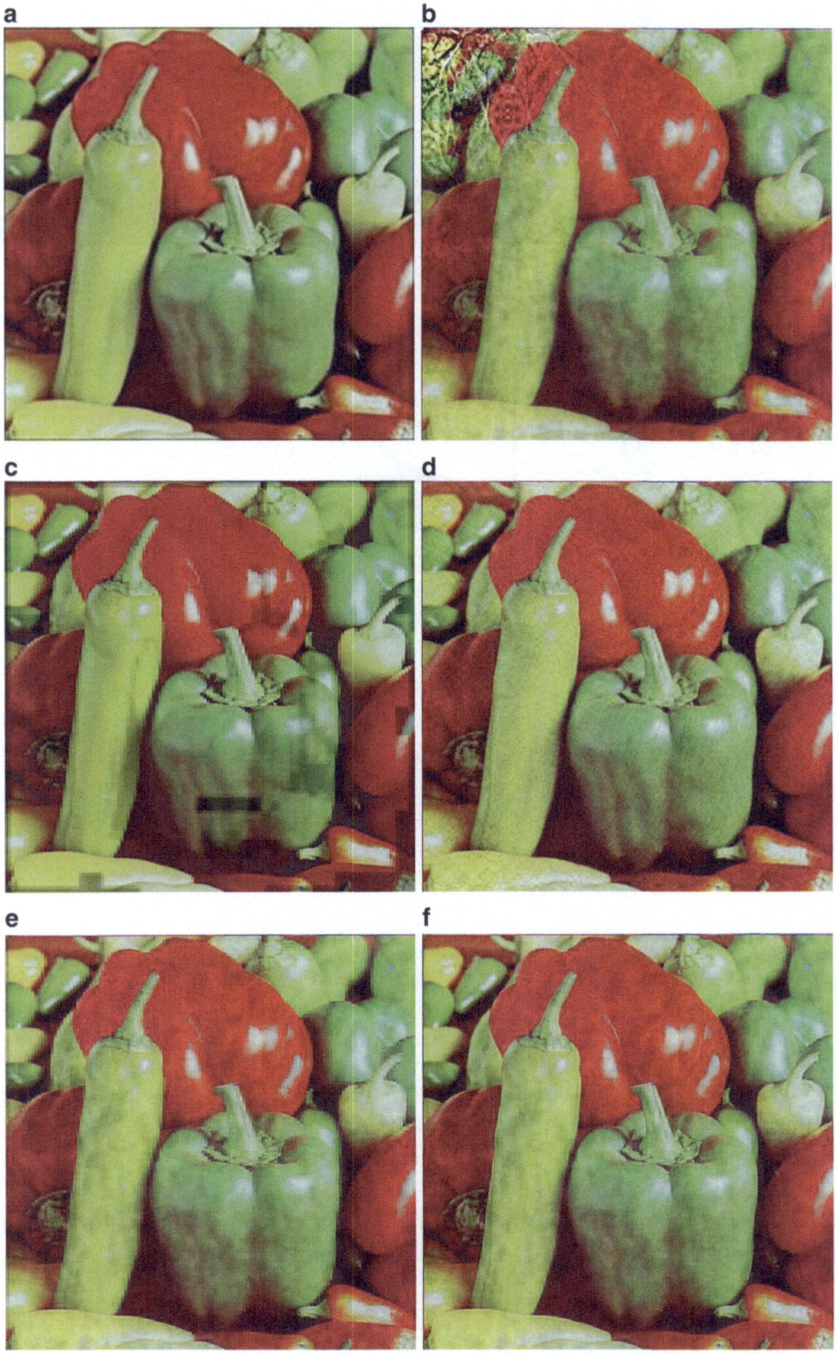

Fig. 8.2 Results of image compression: (**a**) original image; (**b**) compressed image using direct cosine transform (DCT); (**c**) compressed image using discrete wavelet transform (DWT); (**d**) compressed image using singular value decomposition (SVD); (**e**) compressed image using DCT + DWT; and (**f**) compressed image using DCT + DWT + SVD

Fig. 8.3 Fog removal: (**a**) fogged image; (**b**) enhanced image; (**c**) gamma-corrected image

Fig. 8.4 Template matching: (**a**) original image; (**b**) template image; and (**c**) template-matched image

corr2. The entire image is divided into subtypes with an object presented in it. Each image with this object is compared with the template image using a correlator. If the result of the correlator is high for a particular image with the object in the template, then this object is the best matched with the template and is highlighted in the image. A simple example of template matching is shown in Fig. 8.4, in which the number 2 is identified in an enrollment number entered by a student on an answer sheet.

8.6 Image Mosaicing

Image mosaicing is a process in which two images that have been acquired using a camera with a limited capture area are combined [11]. This application can also combine images taken with different resolutions and different lens angles. This application is widely used in mobile phones to capture an image with a wider area. The algorithm for this application can be designed and implemented using a basic image addition process and thresholding. Figure 8.5 shows the application of image mosaicing in which two areas of an institute are acquired and one image is generated showing details from both images.

Fig. 8.5 Image mosaicing: (a) left image; (b) right image; and (c) mosaiced image

Bibliography

1. Thanki, R. M., & Kothari, A. M. (2016). Digital watermarking: Technical art of hiding. *Intelligent Analysis of Multimedia Information, 431*, 426–461.
2. Cox, I., Miller, M., Bloom, J., Fridrich, J., & Kalker, T. (2007). *Digital watermarking and steganography*. USA: Morgan Kaufmann.
3. Honsinger, C. (2002). Digital watermarking. *Journal of Electronic Imaging, 11*(3), 414.
4. Podilchuk, C. I., & Delp, E. J. (2001). Digital watermarking: Algorithms and applications. *IEEE Signal Processing Magazine, 18*(4), 33–46.
5. Chan, C., & Cheng, L. (2004). Hiding data in images by simple LSB substitution. *Pattern Recognition, 37*(3), 469–474. https://doi.org/10.1016/j.patcog.2003.08.007.
6. Shum, H. Y., Kang, S. B., & Chan, S. C. (2003). Survey of image-based representations and compression techniques. *IEEE Transactions on Circuits and Systems for Video Technology, 13* (11), 1020–1037.
7. Gonzalez, R., & Woods, R. (2008). *Digital image processing*. Delhi: Pearson Education India.

8. Wallace, G. K. (1992). The JPEG still picture compression standard. *IEEE Transactions on Consumer Electronics, 38*(1), xviii–xxxiv.
9. Marcellin, M. W., Gormish, M. J., Bilgin, A.,& Boliek, M. P. (2000). An overview of JPEG-2000. In *Data Compression Conference, 2000. Proceedings. DCC 2000* (pp. 523–541). IEEE.
10. Lewis, J. P. (1995, May). Fast template matching. *Vision Interface, 95*(120123), 15–19.
11. Capel, D. (2004). Image mosaicing. In *Image Mosaicing and super-resolution* (pp. 47–79). London: Springer.

Index

R. M. Thanki, A. M. Kothari, *Digital Image Processing using SCILAB*,
https://doi.org/10.1007/978-3-319-89533-8

The manufacturer's authorised representative in the EU is Springer Nature Customer Service Centre GmbH, Europaplatz 3, 69115 Heidelberg, Germany. If you have any concerns regarding our products, please contact ProductSafety@springernature.com

Printed and bound by CPI Group (UK) Ltd, Croydon, CR0 4YY
25/09/2026
02222891-0018